Communication and "The Good Life"

international
communication
association

Annual Conference Theme Book Series

Vol. 2

The ICA Annual Conference Theme Book Series
is part of the Peter Lang Media and Communication list.
Every volume is peer reviewed and meets
the highest quality standards for content and production.

PETER LANG
New York • Bern • Frankfurt • Berlin
Brussels • Vienna • Oxford • Warsaw

Communication and "The Good Life"

EDITED BY
Hua Wang

PETER LANG
New York • Bern • Frankfurt • Berlin
Brussels • Vienna • Oxford • Warsaw

Library of Congress Cataloging-in-Publication Data

Communication and "the good life" / edited by Hua Wang.
pages cm. — (ICA annual conference theme book series; v. 2)
Includes bibliographical references and index.
1. Quality of life. 2. Information technology—Social aspects. 3. Social media.
4. Communication—Social aspects. I. Wang, Hua, editor.
HN25.C636   303.48'33—dc23   2015000941
ISBN 978-1-4331-2856-1 (hardcover)
ISBN 978-1-4331-2855-4 (paperback)
ISBN 978-1-4539-1539-4 (e-book)
ISSN 2330-4529 (print)
ISSN 2330-4537 (online)

Bibliographic information published by **Die Deutsche Nationalbibliothek.**
Die Deutsche Nationalbibliothek lists this publication in the "Deutsche
Nationalbibliografie"; detailed bibliographic data are available
on the Internet at http://dnb.d-nb.de/.

Cover image and concept by Stacey Spiegel
Cover design by Daniela Franz

© 2015 International Communication Association
Peter Lang Publishing, Inc., New York
29 Broadway, 18th floor, New York, NY 10006
www.peterlang.com

Table of Contents

Part II Perceptions, Connections, and Protection

Part III Challenges, Opportunities, and Transformation

Acknowledgments

I am deeply honored to serve as the 2014 ICA Conference Theme Chair and the editor of this theme book. These were no small tasks, as I discovered in the last 24 months. From the conference planning that began even before London to organizing the theme sessions in Seattle, from manuscript reviews to the editorial process, every step of the way presented unique challenges and pushed me to new limits. The learning was tremendous because the people I worked with were inspiring. I would like to thank the ICA President Peter Vorderer for his vision and trust in me with these important tasks. I am also grateful to the ICA Executive Director Michael L. Haley, the 2012 ICA Conference Theme Chair Patricia Moy, and the 2013 ICA Conference Theme Chair Leah A. Lievrouw for sharing their wisdom with me along this journey. In addition, I thank Stacey Spiegel and Daniela Franz for designing the beautiful book cover with the generous support of Sam Luna. Mary Savigar, our dedicated commissioning editor at Peter Lang, has been enthusiastic, understanding, and patient. This book would not have been possible without Mary's expert advice and the professional service by her colleagues Phyllis Korper, Sophie Appel, Catherine Tung, and Suzie Tibor. I also appreciate the encouragement and support from family, friends, as well as colleagues in my home department at the University at Buffalo, especially my Department Chair Tom Feeley. I also want to thank Amanda Hamilton for facilitating the contributor agreements and Vivian

Wu for helping with indexing. Last but not least, I am thankful to all of the contributors in this volume, who are well represented in terms of gender, international background, academic seniority, and scholarly tradition, for the diversity and insight they brought to this volume about communication and "the good life."

Foreword

PETER VORDERER
UNIVERSITY OF MANNHEIM, GERMANY
INTERNATIONAL COMMUNICATION ASSOCIATION
PRESIDENT (2014-2015)

The title of this book—*Communication and "The Good Life"*—was the theme for the International Communication Association's 64th annual conference, which took place in May of 2014 in Seattle, Washington. According to its website, "ICA is an academic association for scholars interested in the study, teaching and application of all aspects of human and mediated communication." But what does this have to do with "the good life"?

The notion of "the good life" has been a topic of inquiry, particularly in philosophy, for a very long time—basically, since the very beginnings of the scholarly tradition. What is a good life, how can it be achieved, and what keeps humans (presumably the only species capable of thinking about such a question) from achieving it? These have been some of the most pressing and enduring questions throughout history.

In contrast, communication is a very young discipline, and it has particularly flourished over the past few decades as our ways of communicating and interacting with one another—with or without electronic media—have changed dramatically within a rather short period of time. Many of the technological advancements that we have witnessed in the past 20 years or so were introduced and offered to us with a promise: The promise was simply that these new developments would make our lives better. Of course, it is certainly not novel to correlate happiness with consumerism; the insinuation that things would generally be better if we owned and used some specific item has long been a core tactic of advertising. Television, for example, was championed as a way to make our lives more interesting by giving us access to the worlds outside our immediate reach. Nonetheless, the magnitude of the more recent incarnation of the promise—which specifically relates to digital and social media—is, in my view, substantively different: It taps into at least two different human aspirations that have always been considered decidedly utopian. First, it offers us access to whatever kinds of information we might desire, anywhere and anytime. Who would not want this sort of omniscience? Second, it assures us that we never have to be alone anymore, no matter where we are. Both pledges speak to fundamental human needs, promising to satiate our most essential desires in ways that have never before been possible. I believe that the unprecedented success of social media has to do primarily with the fact that these promises have been made, they have been largely believed, and (at least to some extent) they have also been kept.

At the same time, other needs, wishes, hopes, dreams, aspirations, and even simple preferences have been compromised or neglected as we have pursued this promised omniscience and ever-present companionship. The mediated world we live in today is indeed Janiform: It in many ways brings us closer to some version of "the good life" while, at the same time, leading us away from it. Although we no longer have to be alone, we also forget how to enjoy solitude. Similarly, we can know and learn almost anything instantaneously, yet as a result we may no longer remember how enjoyable it is to understand something slowly, through consistent contemplation over time.

Moreover, the affordances of these new technologies seem to have shifted, for many, from an opportunity to an obligation: We may now feel that, rather than choosing when and where to be connected to these larger networks of information and acquaintances, we must be permanently available, thus losing the luxury of controlling our time and attention.

In my view, this situation calls for communication scholars to study more thoroughly the opportunities and the dangers, the chances and the risks, that digital media pose in our quest for some version of the good life. This book illuminates the complexity of our modern era; my hope is that it will help us understand how we have come to this point but also inform us as we attempt to predict—and decide—where we will be going next.

Introduction

HUA WANG
UNIVERSITY AT BUFFALO, THE STATE UNIVERSITY
OF NEW YORK, USA
INTERNATIONAL COMMUNICATION ASSOCIATION
CONFERENCE THEME CHAIR (2014)

The ways in which we communicate have been evolving significantly in recent years, in part due to the rapid advancement of technologies. These developments present us new opportunities as well as challenges. As we embrace and celebrate the changes in our environment and our own practice, we also need to reflect on how such changes serve our individual well-being as well as the communities, organizations, and societies we belong to.

The 2014 ICA conference theme—*Communication and "The Good Life"*—was provocative, as the president had envisioned. In the call for papers, we asked: What might a "good life" look like in a contemporary, digital, network society? How might we strike the balance and accomplish that? We received an enthusiastic response from the ICA community with a phenomenal number of submissions and organized theme sessions. Scholars from both social scientific and humanistic traditions participated in stimulating discussions, shared diverse perspectives, and wove together different threads of communication scholarship in our field to better understand this critical moment in human history.

The conference took place in Seattle, a location that fits the theme perfectly. Sitting on the edge of the Pacific Ocean, Seattle has been known for its force of energy through technological and social innovations literally exploding in the city. From coffee shops, edgy restaurants, to sustainable fisheries, urban greening initiatives, and the public art movement, the impact of all these developments have given Seattle the reputation of being in search of "the good life."

In order to symbolically represent such a force of energy and change required to achieve "the good life," we used the artwork of one of this book's contributors as the book cover. The original image, before digitization, was created by lowering a large piece of heavy etching paper just enough to touch the surface of a tank of heated water containing different viscosities of ink mixed together. Therefore, the image itself is a unique and organic print of hydrodynamic energy forces of the water in motion at that particular moment in time. At the micro level, we see the actual force of the water and dynamic expression by the ink. Metaphorically, one might think about inspiration at an epiphany moment feeling a force passing through. At the macro level, we see the tension between various sources of energy and force represented by different shapes and shades, generating an overall impression of explosion. Metaphorically, one might think about the power of change emerged from intense human and mediated communication in the everyday realities of contemporary societies and reflected in this volume.

This book is structured in three parts. The first part focuses on Meaning, Happiness, and Flourishing. Chapter 1, "In Search of the Good Life," uses the framework of exploration to describe and analyze the design of advanced media experiences, connecting the physical world with the virtual world and pointing out the potential of sophisticated technological tools for deeper learning, self-reflection, and personal growth. Chapter 2, "The Good Life: Selfhood and Virtue Ethics in the Digital Age," brings us to the fundamental and philosophical question of what a human being ought to do, or what virtues are required, in order to achieve a life fulfilled with meaning and contentment; it offers us insights into the normative approach in communication research on relational selfhood and digital virtue ethnics by connecting Information and Computing Ethnics and Media and Communication Studies. Chapter 3, "Eudaimonia: Mobile Communication and Social Flourishing," provides a global perspective on the social impact of mobile communication in different countries and population groups and through these examples demonstrates the complex constraints and empowering potential these devices bring to the pursuit of human flourishing. Chapter 4, "Meaningfulness and Entertainment: Fiction and Reality in the Land of Evolving Technologies," broadens our understanding of entertainment media and experiences by identifying their eudaimonic functions as an important area of research, particularly related to interactivity, elevation, and morality. Chapter 5, "Media Policy for Happiness: A Case Study of Bhutan," showcases the efforts put forward by the Bhutanese government in developing and evaluating the National Happiness Index and how media policy can be oriented toward achieving happiness and well-being as a nation-state.

The second part of this book focuses on Perceptions, Connections, and Protection. Chapter 6, "Communication and Perceptions of the Quality of Life," presents 30 years of survey studies on the quality of life perceptions, including the influence of media use on such perceptions and their samples ranging from the American national population to regional metro areas and suburbs as well as college students. Chapter 7, "Tuning in versus Zoning out: The Role of Ego Depletion in Selective Exposure to Challenging Media," reports two experiments on how ego-depletion may affect media users' choice in terms of their preference for cognitively and affectively challenging content. Chapter 8, "The Secret of Happiness: Social Capital, Trait Self-Esteem, and Subjective Well-Being," shows the different relationships that bonding and bridging social capital have with subjective well-being and the role that trait self-esteem plays in these relationships. Chapter 9, "Modeling Communication in a Research Network: Implications for the Good Networked Life," exemplifies the dynamic models of a Canadian multi-disciplinary, multi-institutional network of scholars through their changing connections of knowing, friending, advising, and working together over time. Chapter 10, "Communicating Online Safety: Protecting Our Good Life on the Net," advocates a reintegration of online safety research with the protection motivation theoretical paradigm and offers empirical evidence along with reflections on practical implications.

The third part of this book focuses on Challenges, Opportunities, and Transformation. Chapter 11, "Communicative Figurations of the Good Life: Ambivalences of the Mediatization of Homelessness and Transnational Migrant Families," proposes a framework of communicative figurations in adopting process sociology to explore the lived experiences of homeless people and transnational migrant families in a mediatized world. Chapter 12, "Reimagining the Good Life with Disability: Communication, New Technology, and Humane Connections," challenges the conventional definition of "the good life" and urges us to expand our perspectives with the more enabling and humane features that new technologies afford for people with disability. Chapter 13, "The 20th Anniversary of the Digital Divide: Challenges and Opportunities for Communication and the Good Life," reflects on the development of two decades of research on the digital divide and demonstrates how qualitative methods can offer richer explanations to complement statistics and inform public policies. Chapter 14, "Liberating Structures: Engaging Everyone to Build a Good Life Together," explicates a set of simple yet powerful methods called Liberating Structures and provides concrete examples to illustrate how exactly, without necessarily employing any technologies, one can facilitate group communication by effectively engaging everyone in the process to arrive at surprisingly liberating experiences and outcomes.

Throughout the book, we see that, being immersed in a media saturated environment, we have high expectations of new and emergent technologies. We dream about revolutionary changes that these technologies bring to our lives and how they will instantly produce a better reality for us. But we know there is never a one-sided story. These technologies provide a platform for communication. The use of a particular technology is *one* of many ways we can spend time with ourselves or with others. An important question here though is whether the choices are made and the actions are taken with one's consciousness that connects back to life's purpose.

Having a few thousand Facebook friends and broadcasting the so-called "good life" pictures may only serve as an illusionary self-affirmation and ego-boosting mechanism. A connection is not a relationship. A connection creates possibilities. But a relationship requires efforts and mindfulness to bring real benefit and meaning. The problem is not the technology but how we are using it. Can we make deliberate decisions in terms of when, where, how long, and for what purpose to use these technologies rather than being controlled by an invisible audience and being on a leash without even knowing it?

When we are carried away by habitual behaviors to constantly check the smart phone for text messages and social media updates, we are carried away by feeding our own egos and lose the sight of our real presence. That is when we have rooms filled with bodies but empty minds. And it gets worse when people cannot leave any space for themselves but are only running from one place to the next. Then the bodies become the transport of empty heads. That is why we are all very busy, stressed out, but at the end of the day, it feels like something is missing. Occasionally, we hear ourselves asking: What am I doing? On the other hand, if a text message or a Facebook post is sent with a clear purpose when you are right there with what is happening in life, then these technological platforms facilitate your capacity of being in the present moment and living a meaningful life.

Technologies can serve as a double-edged sword. They can serve as an easy escape from the real challenges in the real word. They also have the potential to help create possibilities for alternative realities, realities that may bring the audience to an awareness of themselves and others, to provide them opportunities to ask deeper questions about their being in this world, and how they might make a better decision or a positive change. Although this book does not offer answers to all the questions asked or solutions to all the problems raised, it sets a good starting point to establish a deeper understanding of our lived everyday experiences in a digital, networked, contemporary society with profound conceptual frameworks, rigorous empirical analyses, and intriguing future directions.

PART I

Meaning, Happiness, and Flourishing

CHAPTER ONE

In Search of the Good Life

STACEY SPIEGEL AND HELI TUOMI CARLILE
PARALLEL WORLD LABS, CANADA

What exactly is "the good life"?

On a mid-afternoon in January, we were in the Arctic Circle, gazing over a frozen fjord. It was pitch dark; the silence was deafening. Only a few dim lights glowing in the distance hinted at human habitation.

This was the Lofoten archipelago in Norway, 68.3333° N, 14.6667° E.

For thousands of years, people had come here in search of "the good life." Why to this dark and remote spot? What on earth made this place so special?

We had been invited as experience design experts to help conceptualize a visitor center that would attract people from around the world. The theme was to be cod.

At first glance, fish and a sparsely populated arctic region seemed to lack the universal appeal necessary for developing a significant international audience. Yet a quick peek into the past revealed a completely different picture: Archaeological evidence suggested that Lofoten had a rich history of human habitation, industry, and trade around fishing, that spanned thousands of years starting in the Stone Age. International trade flourished and brought prosperity to the region. The place boomed, indeed, there was a bona fide "fish gold rush." The fish, however, were not the full story of Lofoten's success. In fact, a large part of their success was also due to the vibrant economic connections and intercultural dialogues facilitated by the latest inventions.

These Norwegians were restless, outward bound, and keen to discover and conquer new worlds. Poised on the edge of a continent and seemingly all coastlines, Norway is a country known for its tradition of exploration. This small nation has produced numerous world-famous explorers, including Fridtjof Nansen, a pioneering North Pole explorer; Roald Amundsen, the first to have travelled to both the North and South Poles; and Børge Ousland, the record maker who crossed the Antarctic alone.

Edmund Hillary once famously said: "It is not the mountain we conquer but ourselves." Thus the mountain (or polar region or ocean), as substantial as it may be, is significant to us only insofar as it is an instrument for *self-exploration* and *personal growth*.

Our goal of this chapter is to discuss the role of technologies through the framework of *exploration*, in the past, the present, and perhaps into the future, in our pursuit of "the good life." From Norway, we trace back to Europe's Age of Exploration and reflect on some of the major inventions in recent human history. Although new technologies are often used to discover the natural world of wonders, we argue they play an even more critical role in our self-exploration and understanding of how we are connected to the external world. Based on the creative projects Parallel World Labs has developed in the past 20 years, we present a multi-lens conceptual model for self-directed deep learning along with a global community and describe the evolution of this model through examples in various content and geographic areas. Finally, we return to our visions for the Lofoten visitor center and elaborate on how different dimensions of "the good life" may all come together through an experience of exploration facilitated by advanced media.

INTO THE NATURAL WORLD OF WONDERS

Europe experienced the so-called Age of Exploration, or Age of Discovery, from the 15th to the 17th century. It was a game-changer for human civilization, helping to lay the basic groundwork of global advancement in the years to come. But what was the impetus for this exploration to occur when it did? The period came on the heels of the Renaissance. This timing ensured that Europe was flush with a unique convergence of economic and political forces that both arose from, and contributed to, the development of new systems and technologies that could enable nations to explore beyond known boundaries.

A major driver of this passion for exploration was the dream of a good life—of power and economic prosperity through discovery of new lands and trade routes—and to this end, resources were made available for the development of new tools to achieve the goal. New technologies spawned new opportunities for exploration. And this is the case beyond Lofoten, Norway, or for that matter, Europe.

FACILITATED HUMAN FLOURISHING

The pursuit of "the good life" has always been an important drive for humankind, and for several hundred years, this pursuit was carried out through exploration, pioneering, urban development, and technological advancements that would make life better. Vast improvements were made in healthcare, science, and telecommunications. Then came the printing press, electricity, telephones, radio, and television. We explored the moon. We also began to access and connect with each other and the world in new ways.

Four hundred years lapsed between the invention of the printing press (1439) and electricity (1820s); only a century passed before the telephone (1876), radio (1895), and television (1926) appeared. It took about 50 years for us to have the first Internet (1974), and then only a few more decades to publish globally and freely on the World Wide Web (1989). Now we live in a digital and networked society. In 2010, Google CEO Eric Schmidt said that every two days we create as much content as was created from the beginning of civilization until 2003 (Siegler, 2010).

The pattern here is one of exponentially rapid development in our pursuit of "the good life." Now, in the early 21st century—four hundred years after the Age of Exploration—we are experiencing that similar dynamic convergence of economic and political forces that enables human exploration on a major scale. This time, however, the direction is as much inward as outward, and the scale is vastly larger: The exploration is no longer in the exclusive domain of professional explorers, philosophers, and nobility, but in the hands of individual citizens.

Spawned from military and government roots, the Internet and advanced telecommunications technologies have flourished with the help of commercial entities that aim to exploit their economic value, and trickled down to individuals where they have cultivated a global phenomenon in personal exploration. These tools —global networks, handheld devices, and vast seas of information—are now in the hands of the people. We are using them to connect with others, to discover new ways to communicate, and to enable ourselves to be faster, smarter, better. Our exploration for "the good life" is taking us to new dimensions. If television had merely promised to make our lives more interesting by giving us access to worlds outside our geographical reach, networked digital media made that promise much more personal, interactive, and satisfying.

TOOLS FOR EXPLORATION IN THE NEW AGE OF DISCOVERY

ex·plo·ra·tion

ˌekspləˈrāSHən/

noun

The action of traveling in or through an unfamiliar area in order to learn about it.

"Climbing is akin to love. It's hard to explain; we endure pain for the joy that comes with discovering ourselves and the planet."

—CORY RICHARDS, NATIONAL GEOGRAPHIC *ADVENTURE /*
EXPEDITION PHOTOGRAPHER AND ALPINE CLIMBER

The "joy" that Richards alludes to may be a powerful motivator behind the popularity of smartphone uptake worldwide. The exhilaration, be it on a mountain or on a smartphone, comes from that same neurological release of dopamine as we engage in something deeply pleasurable, making a connection beyond ourselves.

Our work at the Parallel World Labs has centered around developing experiences that enable a deeper understanding of what constitutes a good life. Each project explores a new way of connection, using technology to facilitate the users to self-direct their own experiences. We help users identify things that are of interest to them, encourage them to follow their own path of curiosity, offer opportunities for inspiration, and enable them to achieve a deeper understanding of the world as a connected place.

There is evidence that some technology users struggle with a kind of addiction to this ubiquitous connectivity, probing obsessively without engaging at a deeper level (Vorderer & Kohring, 2013). In their pursuit of human connection, of knowing, and of "the good life," the medium usurps the message. This "scrolling culture" is of concern not only for its possible personal and social effects, but also for its impact for content developers. Recognizing the popularity of rapid information scanning over more meaningful experiences, many app- and content-providers now prepare their offering in short format in order to cater to this superficial consumption style. Within this ecosystem of hyper-connectedness and social-media-driven expectations, Parallel World Labs is researching ways to offer "deep learning experiences" for audiences with a shorter-than-ever attention span and developing methods that highlight opportunities for the users to explore the world—to use technology as a means to an end, not an end in itself.

The history of human communication has evolved from an oral tradition to written words, and now interwoven into vast global networks. Technology allows us to search, research, organize, compare, and mash different pieces of information together, at a new speed. The cycles have become much shorter and faster. Yet we often find ourselves swimming if not drowning in a tsunami of information, attempting to make sense of what is consumed in these short and rapid cycles. In fact, at the center of all these actions are our live experiences and technology can help us tell our stories in unique ways. At their core, our stories take us back to the oral traditions, except that now we can pause, rewind and fast-forward, probe deeper, mash, and re-interpret those stories, then share them with billions of others around the world just at the fingertip. Our storytelling tradition continues to evolve from a top-down patriarchal narrative to a multifaceted and co-creative dialogue. In the next section, we present our current experimental tools and methodologies and describe some exemplary projects to illustrate the evolution of a deep learning model (see Figure 1.1, from Spiegel & Hoinkes, 2009).

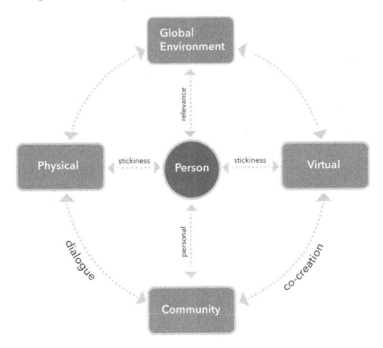

Figure 1.1. Parallel World Labs' Deep Learning Model.

The Multi-Lens Approach

Our central pedagogy emphasizes the capability of technology to enable pluralistic voices and opinions. In designing content and media experiences, we build on a core concept through *multiple lenses*, organizing the subject matter to be viewed from different perspectives, and enable the user to explore through these lenses, especially the viewpoints of others. When used in a museum environment, the Multi-Lens Model is designed to create an associative system for uncovering related rich media content spawned by an artifact. In this way the system feeds a dialogue and inspires curiosity as the visitor decodes the layers of interpretative meaning by viewing an artifact from multiple perspectives.

Parallel World Labs built a prototype mobile app called "The Curiosity Engine" to pilot test the validity of this approach. It was an innovative tool for visitors to explore the meaning behind, context of, and the connection between artists, artworks, and artifacts in a museum exhibition. The Curiosity Engine allowed the visitors to discover the significance of a particular artifact and also helped them develop an understanding of the context for how and why the work itself has gained historical importance. The Curiosity Engine augmented reality, interfacing a wide range of associated media content to stimulate the mind and create a dialogue between the visitor and the artifact.

Modeling and Simulation

We use 3D modeling and animation, particularly within game environments, to create experiences where the user can explore numerous possibilities and make critical decisions using simulations and scenarios. Simulations are a powerful form of exploration of "the good life," offering users a safe way to explore unfamiliar territories and scenarios without jeopardizing the welfare in their realities. For Parallel World Labs, these types of projects are typically seen in the education, training, and healthcare sectors.

Fishing and Oil Industries: Online 3D entertainment worlds aside, "serious" sectors such as fishing and oil industries are also exploring the future what if's— what would it look like if oil and fishing co-existed for their mutual benefit? We are working with the Norwegian fishing industry, oil industry, and The Lofoten Aquarium to envision a new interactive simulation wall within the aquarium for visitors to play and explore the possibilities for co-existence. Using gesture-based navigation on a curved, 6.5-foot tall, and three-panel wall screen, visitors will be able to "play God" by adjusting the level and quality of natural resources, turning the weather engine, and tweaking other forces such as politics, community, and economy to see how each unique combination can affect the harmony or conflict of the fishing, oil, and other coastal industries.

Global Sports: In the world of sport, a game-based learning tool called "Play Fair" was developed for the International Olympic Committee in 2010 and has since been in use at the Youth Olympic Games' Culture and Education Programme. In this case, the specification was to develop an online game for teen athletes who were at risk, during their future careers, of encountering ethical issues around integrity and betting. Parallel World Labs designed a 3D-style Flash game featuring simulated scenarios, whereby the player is asked to make a series of choices around ethical issues within their chosen sport, leading to various outcomes and consequences. The goal of the game was to model possible realities for these users, and use simple game-style simulations to walk them through the decision-making process and educate them about the power of choices and self-improvement.

Social Health and Elder Care: The A2E2 Adaptive Ambient Empowerment of the Elderly project was an EU funded collaboration with AMSTA (The Netherlands), Mawell Ltd. (Finland), Hospital IT AS (Norway), VTT-Technical Research Centre of Finland, and the Vrije Universiteit Amsterdam and Center for Advanced Media Research Amsterdam (The Netherlands). The intent was to focus on quality-of-life improvement for seniors living independently and to specifically support lifestyle changes, such as exercise, sleeping, and healthy eating, which are essential for the prevention and management of diabetes type II and cardiovascular diseases. The project involved connecting a smart environment with a real-time avatar interface as a motivational tool. The animated 3D avatar was equipped with emotional capabilities to act upon the users' individual preferences and respond to their needs.

Culture and Education: In 2007, the Norwegian Ministry of Culture engaged Parallel World Labs to conceive and develop a state-of-the-art experience center for Norwegian music, specifically from the 1950s to the present time. Based on our Deep Learning Model, we designed Rockheim: The National Experience Center of Norwegian Pop and Rock as an immersive experience that integrated rich media assets with educational content in an interactive social environment. Our design of Rockheim is founded in the belief that at its core, musical expression is a powerful tool in defining cultural identity as it evolves from generation to generation. The layered aspect of this experience model enabled people to gain knowledge through a series of intimate immersive experiences that combine real world artifacts with layers of related virtual content, inviting exploration through a multi-lens approach. Visitors can explore various aspects of content that triggers their particular interest, then compare and contrast them in ways that could not have been achieved by using traditional didactic models of communication. In this learning model, technology supports self-directed exploration of a world filled with richer meaning and a deeper understanding.

Healthcare: A massive two-story interactive LED wall, featuring multi-user interactive video storytelling, greets hospital visitors in the main atrium of the St. Olavs Knowledge Center in Norway. The narrative storylines in this public health communication tool were designed to improve the quality of life for patients and their families. The experience offers hundreds of short, self-directed educational videos on current medical procedures that manifest and contextualize the innovative work taking place at this top research hospital in Europe. One of the more recent interactive storylines highlights the work of the Mosers—the resident Nobel prize-winning team engaged in brain research.

RETURN TO 5,000 YEARS OF THE GOOD LIFE

Our multi-lens approach and deep learning model has evolved over the last two decades. As we were standing in the pitch dark, overlooking at the reflections of the lights on the frozen fjord in this January mid-afternoon, and pondering the history, economy, culture, technology, and humanity, it all came together in our visions for the new international visitor center in the Lofoten archipelago of Norway. The goal is to create a place where people locally, regionally, nationally, and internationally can share in a dialogue and exploration of what constitutes "the good life."

For the past 18 years, a community now known as Storvågan in Lofoten had been having an ongoing discussion about how to tell their unique story to the half million visitors to the region each year. Storvågan, whose history of Stone Age settlements date back at least 5,000 years, sits a few kilometers outside of Kabelvåg, a former fishing village. For thousands of years, people living in this area had a recurring story to tell regarding the return of the skrei. Skrei is a type of cod that swims in massive numbers from the Barents Sea to spawn each spring in Lofoten, making this cod fishery one of the world's oldest and largest.

Now, at a time when the value of ecological tourism has become of higher commercial value than the skrei fishery, a new complicating factor entered the picture of this utopian environment: the discovery of oil deep below the seabed. For a community that has based its survival on the managed exploitation of its ecological resources, the discovery of oil should not have been a cause for concern. However, the discovery and the ensuing discussion of its exploitation have rapidly become hot topics of discussion and political alliance.

The divisive nature of the "eco vs. oil" discussion seems to have been a motivating factor in bringing in an outsider to envision a new, compelling, and united strategy for telling the story of skrei. The project goals were twofold. For stakeholders, economic development through a new tourism attraction was essential, but a second important goal was to help communicate to local youth a renewed vision of their history, culture, and identity, the loss of which in recent generations had become increasingly obvious. Young people were leaving the region in droves.

The skrei project needed to find effective tools to aggregate and communicate the identity and culture of a community, while framing it in a way that would be relevant to local, regional, national, and international audiences. The cod narrative would require an innovative approach. While the leaders of the community wanted to highlight the annual return of the skrei as the iconic symbol of the community's existence, we perceived a more universal meaning.

Parallel World Labs began by speaking with locals and examining the region's history. All of the conversations had one common thread: People chose to be here, in Lofoten, because they felt "it was a good life." Despite the long dark winters and remote arctic location, there was a plenitude of natural resources and stunning scenery. But the deeper reasons lay in how the geography played out in the human psyche: The unique combination of natural resources and setting—a long north-south coastline—had created a unique catalyst for exploration and trade thanks to innovations that had spurred advancement in boating, fishing, and oil detection technology, as well as transportation and trade, arts and culture, telecommunications, and politics.

Skrei is what sustained families, grew communities, and built cities. It not only fed the people, but also brought them wealth: a renewable resource, a transportable commodity, and an international currency. Thanks to skrei, Lofoteners had the resources to develop sustainable industries, vibrant international markets, and trade routes that persist even today. While cod fisheries elsewhere in the world have been depleted and bankrupted, Lofoten's industry continues to flourish.

Skrei, we perceived, presented a unique symbol for locals and visitors alike to contemplate the meaning of "the good life": what it is, how we create it, and how we sustain it. For the local community, skrei offered a vehicle for interpreting past, present, and future, and a renewed perspective of Lofoten's role in national history and success.

Suddenly, we began to see the global attraction for a Lofoten experience center: Here was a place that embodied, in one unique set of aspects, "the good life." Although it was only one example, it was a robust example worth sharing, exploring, and discussing. Bringing a global multicultural lens to Lofoten might yield something worthwhile, and it could happen both through on-site immersive experiences and events, and also virtually by facilitating online forums on sustainable design and development. Participants could explore social and political issues, and think about them in the context of their own circumstances whether at a personal, regional, or national level.

What emerged was a vision to build the SKREI Opplevelsessenter [Experience Center]: In Search of the Good Life, which would establish in Lofoten a state-of-the-art cultural destination and international research center (see Figure 1.2). The SKREI Experience Center, owned by Museum Nord and Storvågan AS, would offer an immersive self-guided journey through the story of skrei, their annual return to Lofoten, the human culture built on it, and how it weaved through every aspect of the local experience. Next door, the Research Center would act as a global think tank, facilitating conferences and events to foster dialogues on sustainable development, and strategies and techniques for building a good life within global populations. The complex would also feature a test kitchen facility for offering Lofoten culinary art workshops on cooking with skrei, as well as other classes and events to bolster food tourism. The SKREI Experience Center, still in planning as of 2015, would aim to create a unique meeting place for dialogue on a critical global topic.

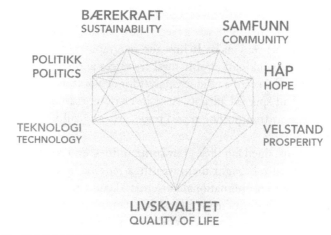

Figure 1.2. The "DIAMOND" Thematic Foundations of the Center for the Good Life in Lofoten.

CONCLUSION

In this new Age of Exploration, our capacity to be globally networked, connected by personal devices, sensors, and data feeds, will be the vessels that convey us to both new worlds and new experiences of old worlds. However, our newfound ability to connect physical and virtual worlds is merely the latest iteration in our search for meaning, connection, and "the good life." We are still in our infancy in some aspect: As new tools are created, new methods and experiences become possible. We can only speculate as to what new worlds and dimensions we will find as our vehicles of exploration become more sophisticated. For the time being, our hand-held devices have become our personal compasses, helping each of us locate and navigate ourselves within the sea of human experiences. As the old adage suggests, if the journey is more vital than the goal, then our evolving yet relentless exploration may in itself be "the good life."

REFERENCES

Siegler, M. (2010, August 4). *Eric Schmidt: Every 2 days we create as much information as we did up to 2003*. Retrieved from http://www.techcrunch.com/2010/08/04/schmidt-data

Spiegel, S., & Hoinkes, R. (2009). Immersive serious games for large scale multiplayer dialogue and cocreation. In U. Rittefeld, M. Cody, & P. Vorderer (Eds.), *Serious games: Mechanisms and effects* (pp. 469–485). New York: Routledge.

Vorderer, P., & Kohring, M. (2013). Permanently online: A challenge for media and communication research. *International Journal of Communication, 7*, 188–196.

CHAPTER TWO

The Good Life

Selfhood and Virtue Ethics in the Digital Age

CHARLES M. ESS
UNIVERSITY OF OSLO, NORWAY

"Perhaps it would not be a bad idea for the teams at present creating cybernetics to add to their cadre of technicians, who have come from all horizons of science, some serious anthropologists, and perhaps a philosopher who has some curiosity as to world matters."
—Père Dubarle, cited in Wiener (1950/1954, p. 180)

I take the conference theme for this year's ICA—Communication and "the good life"—as a high watermark of a growing interest in the field of communication research in overtly *normative* approaches. "The good life" is a core focus and hallmark concern of virtue ethics, so I begin this chapter with a brief introduction to virtue ethics. We will see that virtue ethics is well grounded in such ancient figures as Plato and Confucius, and re-emerges in both early and later modernity. I focus on these more recent appearances of virtue ethics—first, within the disciplines of Information and Computing Ethics (ICE) and then within Media and Communication Studies (MCS): In between, I review my own contributions along these lines as a "bridge worker" between these two domains. I argue that these appearances trace out a growing cross-disciplinary interest in appropriating virtue ethics as part of a larger pattern of projects in otherwise largely "value-free" or "value-neutral" disciplines that in fact aim to make *normative* claims. These claims include precisely what constitutes "the good life" as a life we *ought* to pursue and as a life characterized by certain normative *values* such as human contentment and flourishing. To be sure, normative approaches within the social sciences in general and MCS in particular are not entirely novel. But these recent developments are much more radically cross-disciplinary than before, including emerging bridges between virtue ethics and ICE on the one hand, and the empirically-oriented approaches of MCS on the other. Finally, I close with suggestions for how such cross-disciplinary work may offer us promising new approaches to discerning what good lives in our contemporary societies, as increasingly shaped and defined by information and computing technologies (ICTs), might look like and what sorts of concrete steps we can take to pursue these.

VIRTUE ETHICS: A BRIEF INTRODUCTION

As I have described more fully elsewhere (Ess, 2013), virtue ethics emerges among the most ancient ethical and philosophical traditions in both the East and the West (e.g., Buddhism, Hindu thought, Confucian traditions, as well as the Abrahamic religions and ancient Greek sources such as Plato and Aristotle). This global ubiquity may be due in part to what I think of as a both empirical and common-sensical approach to life in virtue ethics, one defined by what seems to be an unavoidable human question:

> What must I (learn to) do—what abilities, practices, habits (*virtues*) are required—in order for me to achieve a life marked by contentment or deep happiness (*eudaimonia* in ancient Greek)?

While historically the English word "virtue" has translated the Greek *arête*—*arête* is better translated as "excellence." As computer ethicist Deborah Johnson (2001) puts it, "...ethics was concerned with excellences of human character. A person possessing such qualities exhibited the excellences of human goodness. To have these qualities is to function well as a human being" (p. 51).

This sense of *eudaimonia* is bound up with an experience of—including a series of *practices* or efforts to bring about—*harmony*, both internally, within the *psyche* ("soul" or self) of an individual and externally, with our larger social contexts. For our purposes, it is particularly important to note that virtue ethics emerges in these ancient traditions in conjunction with a strongly *relational* sense of selfhood. A relational self is constituted primarily by the relationships that define it, first within the family (e.g., parent-child) and then within our larger communities and societies. This focus on relationality is apparent, for example, in Aristotle's attention to *friendship* as a core component of "the good life" where friendship requires specific excellences such as patience, amiability, honesty, and wit (Vallor, 2010). Such excellences are precisely the virtues that thus contribute to our sense of contentment and harmony. And so, as computer ethicist Deborah Johnson puts it, "...ethics was concerned with excellences of human character. A person possessing such qualities exhibited the excellences of human goodness. To have these qualities is to function well as a human being" (Johnson, 2001, p. 51). Moreover, these capacities do not come naturally: Rather, they are difficult and must be practiced in order to be available for our use more easily. Their development thereby represents the unfolding of our best potentials—and insofar as we succeed in doing so, such unfolding contributes to our sense of deep happiness and flourishing.

Virtue ethics was obscured—but by no means extinguished—by the rise of Christianity; it then came to the foreground again with the Renaissance and the recovery of older ethical traditions. It became especially central in the Modern Enlightenment—i.e., precisely as part of a conscious turn away from religiously rooted norms and sensibilities toward rational ones. Rosalind Hursthouse (1999) has argued that virtue ethics have begun to flourish again as we have come to recognize that for all of their advantages, neither utilitarianism nor deontology take up what we recognize as desirable, if not simply necessary, for a complete moral life: These include "moral wisdom or discernment, friendship and family relationships, a deep concept of happiness, the role of the emotions in our moral life, and the questions of what sort of person I should be" (p. 3).

Furthermore, I have argued that virtue ethics is increasingly attractive precisely because we in the West are becoming increasingly *relational* sorts of selves—i.e., we are shifting away from the strongly *individual* senses of selfhood that root both deontology and utilitarianism (Ess, 2014). These shifts are apparent, for example, in prominent contemporary social science theories that emphasize relational selves in various ways (e.g., Goffman, 1959; Simmel, 1955). Relational selfhood further underlies our clear shift away from *individual* conceptions of privacy toward more relational ones—perhaps most prominently in Helen Nissenbaum's (2010) theory of privacy as contextual integrity (cf. Ess & Fossheim, 2013).

As last examples: Recent feminist philosophy has sought to develop notions of "relational autonomy"—notions that preserve high modern emphases on moral agency and freedom, but as conjoined a resolute insistence that these arise for us as relational beings within and through our social relationships rather than independently of or, in Hobbesian fashion, only in conflict with such relationships (Mackenzie, 2008; Westlund, 2009). These relatively new understandings of relational moral agency, as reinforced by social scientific accounts, are then incorporated as part of a basic philosophical anthropology that undergirds a recent European Commission project that seeks to help us rethink what it means to be human in a digital era. The Onlife Manifesto affirms that the self is an inherently relational [and] free [individual] self (Broadbent et al., 2013). This project is worth noting as it marks out how relational selfhood is becoming increasingly recognized and taken up, and in this instance along with some attention to virtue ethics as well—and all of this at a policy-directed level of the European Union (cf. Simon & Ess, forthcoming, and further comments below).

THE (RE)EMERGENCE OF VIRTUE ETHICS IN CONTEMPORARY DISCIPLINES

Information and Computing Ethics (ICE)

The foundational text of contemporary ICE is Norbert Wiener's (1950/1954) *The Human Use of Human Beings*. Wiener is most widely known for his pioneering work in establishing the field of cybernetics—what he describes as "the theory of messages," and what is more generally understood as the science of self-correcting (information) systems (Wiener, 1950/1954, p. 77). Wiener draws on the ancient Greek term Κυβερνήτησ [cybernetes], meaning a steersman, helmsman, or pilot. But this term, as it appears in Plato, is not about value-neutral informational self-steering: On the contrary, the Κυβερνήτησ is a centrally *normative* figure or image. Plato describes the Κυβερνήτησ this way: An outstanding pilot or doctor is aware of the difference between what is impossible in his art and what is possible, and he attempts the one, and lets the other go; and if, after all, he should still trip up in any way, he is competent to set himself aright (*Republic* 1991, p. 38, 360e–361a). Specifically, for Plato the Κυβερνήτησ is a primary model or example of the ethically *just* human being and ethically *just* ruler. More precisely, the Κυβερνήτησ embodies the critical capacities of being able to discern our possible ethical choices within a specific context, coupled with our ability to make the ethical equivalent of mid-course corrections should we find that our initial judgments and decisions were mistaken. "Cybernetics" in this way is founded on a Platonic notion of *ethical* self-correction.

These roots in Platonic virtue ethics are then part and parcel of Wiener's essayist explorations of what new computational technologies might make possible and thereby what new possibilities of human *flourishing* might emerge. Specifically, such flourishing includes Wiener's affirmation of the intellectual-moral tradition of the Enlightenment and Western liberalism, including the motto of the French Revolution: Liberté, Egalité, Fraternité. Specifically, Wiener (1950/1954) understands "liberty" to mean "…the liberty of each human being to develop in his freedom the full measure of the human possibilities embodied in him" (p. 106). Recalling Aristotle, then, freedom for Wiener is just the freedom to develop our (best) potentials as human beings. In this direction, new computing technologies should be designed and taken up in our lives to foster such freedom, such unfolding, and such human flourishing (cf. Bynum, 2000).

In contemporary ICE research and literature, virtue ethics continues to play a prominent role on both more theoretical and more operational levels. In the more theoretical direction: For more than a decade, Luciano Floridi has unfolded a *philosophy of information* and a correlative *information ethics* that has achieved a defining centrality in ICE. Floridi (2013) draws directly from Wiener's work and incorporates a number of features of virtue ethics in his own information ethics; at the same time, Floridi offers some important critiques of virtue ethics—at least from the standpoint of his project to develop a far more encompassing and comprehensive information ethics based, as he points out, on a very different set of assumptions regarding human beings and the larger world than operate in most versions of virtue ethics.

In a more applied direction, Shannon Vallor (2010, 2012, 2014) has extensively explored and applied virtue ethics to especially the online experiences and potentials of social networking sites (SNSs), thereby making virtue ethics utterly central to the ethical analyses of what are arguably the most important venues of contemporary online experience. Vallor (2010) uses Aristotle's analysis of friendship (as a primary and all but essential component of the good life for human beings as intrinsically social beings) to identify a number of virtues that she argues are key to *communication*—including patience, perseverance, empathy, and trust. Vallor analyzes the design and affordances of contemporary SNSs, arguing that these sites thereby privilege and reinforce short, quick, and easy forms of communication. These venues thus stand in sharp contrast with the offline communication environments that foster the virtues of patience, perseverance, and so on: As we have seen, such virtues, especially at the beginning are difficult to acquire, and so require precisely long, hard practice—whereas the affordances and design of online venues allow us to easily quit and escape. To be sure, Vallor (2012) recognizes that SNSs and other forms of online communication *may* help foster important communicative virtues (reciprocity, empathy, self-knowledge, and the shared life)—virtues that are thereby arguably necessary for leading lives of contentment in a digital era: But they will do so, she argues, only if they are designed to do so (Vallor, 2014).

Virtue Ethics in ICE and MCS

Especially over the past decade or so, I have attempted to develop a number of conceptual bridges between ICE and MCS—precisely in the domain of applied ethics. In this work, virtue ethics has taken an increasingly central role. This work began with the Ethics Guidelines Committee of the Association of Internet Researchers (AoIR), which developed the first published set of ethical guidelines for Internet research (Ess & AoIR, 2002). In subsequent work, including the first edition of

Digital Media Ethics (Ess, 2009), I expanded the application of diverse ethical frameworks—including virtue ethics—to a range of issues in what I called digital media ethics "for the rest of us." That is, in contrast with the more specialized issues that philosophers, computer scientists, and affiliated fields tend to take up (e.g., system security, customer service), I have sought to develop an ethical toolkit to assist all of us who use digital media in attempting to resolve issues we confront every day, such as privacy, copyright, and sexually explicit and/or violent content online.

Virtue ethics was hence an important component of my work up to 2009. But as I prepared to revise *Digital Media Ethics* for its second edition (Ess, 2013) and undertook a fresh look at how these domains and literatures had developed—it was striking to see how virtue ethics had become increasingly central as exemplified in Vallor (2010, 2012, 2014). And as my own work became increasingly oriented toward the practical and the everyday, virtue ethics accordingly became still more important. Most publicly, the European Commission's Onlife Initiative brought together a group of 13 scholars in philosophy, media and communication studies, sociology, anthropology, law, and computer science to first undertake a "conceptual re-engineering exercise" that sought to evaluate and, where necessary, reconfigure and redesign the most fundamental assumptions and concepts that had guided ICT policy and development in the European Union over the past several decades. From here, we developed a policy-oriented set of recommendations in the form of "the Onlife Manifesto." These recommendations include a focus on public and political spaces as constituted by human beings as relational selves whose interactions with one another are all but exclusively mediated by ICTs along with the insistence on the importance of caring "for our attentional capabilities" (Broadbent et al., 2013, p. 10; cf. Simon & Ess, forthcoming). As we have also seen above, such relational selves are consistently affiliated with a virtue ethics approach, therefore, the Onlife Manifesto can be read as a springboard for normative claims—beginning with the injunction to care for our attentional capabilities—that are justified from a virtue ethics approach.

In this direction, we (Ess & Fossheim, 2013) have argued that a good life for such relational selves—including a sense of agency and control over one's life, in a democratic society and social context marked by equality and gender equality—is possible in the digital era: Such a good life, however, will require specific attention to our fostering both "classical" forms of literacy (in Medium-theoretic terms, of *literacy-print*) as these correlate with high modern notions of autonomous (relational) individuals (cf. Wong, 2012). Most recently, I have expanded on these arguments so as to further take on board insights and critiques from mediatization theories in a chapter titled precisely "Selfhood, moral agency, and the good life in mediatized worlds?" (Ess, 2014). This current chapter thus represents still further development of such bridges between philosophical ethics and MCS—a research trajectory that I hope will continue in fruitful ways for some time to come.

In my reading, one of the most significant contributions to MCS literature as invoking virtue ethics is offered by Nick Couldry (2013). Couldry takes as his primary framework a neo-Aristotelian virtue ethics (as articulated by Alisdair MacIntryre, among others) to argue first of all that "media"—most especially the journalistic media—are a *practice* in the sense affiliated with virtue ethics. This means "What we do with media *matters* for how humans flourish overall in an era where we are dependent on the exchange of vast amounts of information through media," which in turn means that "Media ethics in the digital age involves *all of us*, not just media professionals" (Couldry, 2013, p. 25). On this basis, the starting question for media ethics is:

> What are the virtues or stable dispositions likely to contribute to us conducting the practice of [journalistic] media well—"well," that is, by reference both to the specific aims of media as a human practice and to the wider aim of contributing to a flourishing human life together (Couldry, 2013, p. 25).

From here, Couldry (2013) identifies and defines especially three virtues that are critical both to journalists and "the rest of us," namely, accuracy, sincerity, and care.

As a second example: Lee Rainie and Barry Wellman (2012) took up the language of virtue ethics in their book, devoting an entire chapter to "Thriving as a Networked Individual." This includes an "ethical literacy," one focused on building "trust and value...by being *accurate* and thoughtful" with whatever information one distributes through one's networks (Rainie & Wellman, 2012, p. 274; emphasis added). It is important to notice here that the conception of selfhood in play remains within high modern frameworks of individual senses of selfhood and identity. For example, Rainie and Wellman (2012) endorse acting as "autonomous agents" as part of such thriving. As we have seen, a more explicit focus on virtue ethics usually emphasizes a more relational sense of selfhood; but we have also seen that both aspects of selfhood are conjoined in recent accounts of relational selfhood (Broadbent et al., 2013) and relational autonomy (Ess & Fossheim, 2013; Mackenzie, 2008; Westlund, 2009). As a reminder, retaining more individual forms of autonomy is critical to justifying and sustaining democratic politics and norms such as equality and gender equality. In this direction, networked individualism thus highlights an emphasis in selfhood that is key to democratic norms and processes, and thus to "the good life."

CLOSING COMMENTS: META-DISCIPLINARY CONSIDERATIONS

I hope this sketch clearly marks out an emerging trajectory of interest and research in virtue ethics as a primary ethical framework for addressing normative interests in media and communication studies—specifically, new communication technologies' possible roles in and contributions to our widely (if not more or less universally) shared interests in pursuing good lives marked by well-being, harmony, and flourishing. Presuming this trajectory continues to develop and expand, it seems clear that this ineluctably cross-disciplinary work has much to offer us, both within the academy, and, more urgently, to all of us as human beings as such, and as citizens in democratic or democratizing societies. To begin with, precisely as new communication technologies continue to interweave both individuals and larger societies into ever-more interdependent social, economic, and political conjunctions—what is needed in the face of the global diversity of religious and ethical traditions is an ethical framework to guide our common lives and work together. As we have seen, virtue ethics is a primary candidate for such a framework—first of all because it is, so to speak, already embedded and in play in more or less every religious and ethical tradition at work in this wide array of societies and cultures. At the same time, virtue ethics seeks to be resolutely *pluralistic*—i.e., offering ways of our sharing common norms and values (e.g., privacy, community well-being, and so on) that can nonetheless be interpreted and applied in specific cultural contexts precisely so as to preserve local traditions and practices. Indeed, this is not simply theoretical hopes and dreams, but can be seen to emerge as a reality in unfolding ethical practices and norms, such as informational privacy and rights to protection of personally identifiable information (Ess, 2013).

At the same time, however, for such a trajectory to continue to develop and expand, we—i.e., scholars, researchers, and practitioners whose work falls under the two large umbrellas of ICE and MCS—will have to confront and resolve the challenges this trajectory represents to our respective domains and disciplines. Most briefly: We will have to undertake nothing less than a fundamental reconceptualization of our disciplines and the traditional boundaries that have kept them separate over the past century (or more).

That is, at one level, perhaps it is not especially remarkable that such recognized social scientists as Lee Rainie and Barry Wellman can devote an entire chapter in their recent book to suggestions for how to flourish in the digital era. To do so, as we have seen, is not only to invoke the language and trajectories of classical virtue ethics, but to also flirt very closely with offering ethical guidelines—e.g., in the form of "you ought to do _x_ if you want to be happy…" But as the title of ICA 2014 makes clear, Rainie and Wellman are by no means alone in this flirtation. On the contrary, *contra* the generic intention of remaining "value free" or "value neutral" for the sake of a social science that thereby will more closely mimic the so-called hard sciences as *sciences*, as forms of reliable, objective, empirically-grounded knowledge—within MCS there have long been both individuals and schools willing to make their normative foundations and aims clear and explicit. To my knowledge, the most prominent examples are the work of scholars in the fields of critical studies and political economy (e.g., Fuchs, 2014; Lunt & Livingstone, 2012).

At the same time, however, it seems clear that if researchers and scholars in MCS seek to broaden their engagement with philosophically-informed approaches to normative communication—including approaches that explicitly invoke virtue ethics—we will need to revisit and revise the once defining commitments to "value free" and "value neutral" approaches in the social sciences.

Arguably, we are already well on our way—first, broadly, as a range of "post-positivist" approaches to what counts as science have demonstrably undermined the classical models of positivist science, including an ostensibly clear dividing line between the "objective" and the "subjective"—i.e., where the objective was taken to be value-free or value-neutral, condemning moral norms to the domain of merely subjective and thus relativistic and to be set aside. At the same time, a number of contemporary theoretical approaches have taken on board the resulting blurring of the positivist boundaries—including constructivist theories in general, and, for example, in Actor-Network Theory, which attributes agency and norms to non-human actors (cf. Stahl, 2006).

More specifically, I would suggest that we draw important lessons learned from recent such cross-disciplinary projects, beginning with the development of the first and second AoIR ethics guidelines for Internet Research (Ess & AoIR, 2002; Markham & Buchanan, 2012). Inspired in part precisely by the focus on virtue ethics on ethical judgment as embedded in everyday practice, our approach was (and is) exactly to develop research guidelines from "the bottom up," i.e., building first of all on the ethical intuitions and judgments of researchers on the ground and in their field. This contrasts with the more usual approach in many forms of philosophical ethics of attempting to define such guidelines by starting from more general ethical theories, identifying the specific issues and contexts of a given researcher and her ethical challenges, and then somehow deducing guidelines as a specific conclusion. The latter is much easier to do in the comfort of one's philosophical armchair: But our now extensive experience has demonstrated that the results enjoy far less credibility, much less real interest and engagement, among researchers trained more in social science than in philosophy. By contrast, the former approach forces philosophers to re-present philosophical theory and insights in ways that clearly and directly address the ethical intuitions and judgments of researchers where they are. And as a result, the guidelines and insights that become articulate through this highly interdisciplinary process are recognized as relevant, practicable, and *useful*, as is suggested by the broad acceptance and use of the AoIR guidelines among researchers around the globe (Markham & Buchanan, 2012, p. 2).

Inspired by the success of the AoIR guidelines, I would suggest that future collaborations between MCS and ICE might fruitfully proceed by at least initially considering our approaches to such interdisciplinary dialogue and collaboration as one possible model. As a first step, we should take on board Aristotle's admonition in *The Nicomachean Ethics* that "it is the mark of an educated human being to look for precision in each class of things just so far as the nature of the subject admits; it is evidently equally foolish to accept probable reasoning from a mathematician and to demand from a rhetorician scientific proofs" (Book 1, section 3: 1094b). In the terms that we used in the first AoIR guidelines, this means (again) that our ethical analyses and resolutions are ineluctably characterized by ambiguity, uncertainty, and multiple (ideally, pluralistic) ethical *judgments*. Or, most simply, *contra* expectations for single, final, certain "answers" to our ethical challenges, as if ethics were like a mathematical proof or logical deduction—in general, the best we can hope for are guidelines.

As this example suggests—and as reinforced by the overall trajectory I have traced out above—virtue ethics recommends itself as a primary (but by no means the only) ethical framework for future collaborative work. Somewhat more specifically, I would argue that one of the key virtues of virtue ethics is that it brings to the foreground what appears to be a widely, if not universally-shared existential human concern, as it explicitly asks: What might the good life be for (embodied) relational autonomies as citizens in and co-constructors of mediatized societies?

Still more particularly, I would presume that we will continue to define "the good life" in terms of the high modern norms and values that cluster around a (relational) autonomy: Such norms would include justice, fairness, equality (including gender equality) as components liberal-democratic polities. Given this, virtue ethics would immediately provide two directions for further reflection and development. The first would be a *critical* perspective on contemporary communication media and technologies as exemplified above in the work of Shannon Vallor. That is, given a virtue ethics commitment to such foundational norms and values as part of a larger understanding of and orientation toward a good life constituted by well-being, harmony, and flourishing—we can thereby critically and empirically interrogate how far extant media technologies and practices indeed foster and/or frustrate our efforts toward pursuing a good life in a digital era. Indeed, as Vallor has suggested, this critique can turn in a positive direction, so as to guide our choices not only in terms of our consumption and uses of specific technologies and communication venues: More fundamentally, virtue ethics criteria can be taken up in the *design* of new media technologies as well.

Second—and this may well be the most challenging aspect of taking on virtue ethics in a serious way—virtue ethics would bring to the foreground the need for an informed, conscious, on-going cultivation of the self: Cultivation, more specifically, of the virtues—the practices, excellences, habits—requisite for our pursuing the good life in a digital era. As a reminder, Vallor has highlighted the importance of patience, perseverance, and empathy as virtues especially critical to the sorts of communication required for deep friendship, long-term intimate relationships, and so on. For his part, we have seen Couldry argue for the virtues of accuracy, sincerity, and care. Along these lines, on the basis of Medium Theory and the work of Michel Foucault (1988) on writing as a technology of and care for the self, I have argued for the importance of writing and the skills and abilities affiliated with *literacy-print* (as correlated with high modern understandings of selfhood as a rational autonomy, where such autonomy is foundational precisely for democratic practices and norms) alongside fostering the skills and abilities affiliated with the "secondary orality" (Ong, 1988) or "tertiary orality" (Rasmussen, 2014, pp. 26–28) of electric or electronic media (as correlated with more relational forms of self-hood: Ess & Fossheim, 2013).

As a final point, a key virtue that we as scholars and researchers interested in such collaborations will also need to cultivate and practice will be a form of *epistemological humility*. Such humility is needed to counter our tendency—all too common and all too commonly encouraged in the academy—to presume and establish the superiority of our own knowledge and disciplines over against those of "others." To state the obvious: Dialogue and collaboration best proceed and succeed on foundations of equality, respect, and trust. Epistemological arrogance undermines such foundations and thus makes collaboration and dialogue—especially over the long term—all but impossible.

The good news is that the success of such ambitious interdisciplinary collaborations as the development of the AoIR ethical guidelines, the Onlife Manifesto, and many others demonstrates that many of us, at least, are aware of the crucial importance of such humility—and, however contradictory this may sound, may even be reasonably good at practicing it. The pressing news is that such collaborations appear to be increasingly urgent in the face of our growing recognition of the necessity of developing shared norms and ethical practices for all of us whose lives are increasingly interwoven with one another—whatever our cultural traditions and norms (including religious ones)—through our expanding engagements with new media technologies.

In light of this urgent need, and inspired by the initial successes traced out here, I hope this review and reflection will encourage further such collaborations, thereby contributing further to explicitly normative approaches to media and communication technologies.

REFERENCES

Aristotle. (1980). *The Nicomachean ethics.* (W. D. Ross, Trans.). Oxford: Oxford University Press.

Broadbent, S., Dewandre, N., Ess, C., Floridi, L., Ganascia, J.-G., Hildebrandt, M., ...Verbeek, P.-P. (2013). *The onlife initiative.* Brussels: European Commission. Retrieved from https://ec.europa.eu/digital-agenda/sites/digital-agenda/files/Onlife_Initiative.pdf

Bynum, T. (2000). A very short history of computer ethics. *Newsletter on Philosophy and Computers.* Retrieved from http://southernct.edu/organizations/rccs/a-very-short-history-of-computer-ethics

Couldry, N. (2013). Why media ethics still matters. In S. Ward (Ed.), *Global media ethics: Problems and perspectives*, (pp. 13–29). Oxford: Wiley-Blackwell.

Ess, C. (2009). *Digital media ethics*. Oxford: Polity Press.

Ess, C. (2013). *Digital media ethics* (2nd ed.). Oxford: Polity Press.

Ess, C. (2014). Selfhood, moral agency, and the good life in mediatized worlds? Perspectives from medium theory and philosophy. In K. Lundby (Ed.), *Mediatization of communication* (Vol. 21, *Handbook of communication science*, pp. 617–640). Berlin: De Gruyter Mouton.

Ess, C., & the AoIR ethics working committee. (2002). Ethical decision-making and Internet research: Recommendations from the AoIR ethics working committee. Retrieved from aoir.org/reports/ethics.pdf

Ess, C., & Fossheim, H. (2013). Personal data: Changing selves, changing privacies. In M. Hildebrandt, K. O'Hara, & M. Waidner (Eds.), *The digital enlightenment yearbook 2013: The value of personal data* (pp. 40–55). Amsterdam: IOS.

Floridi, L. (2013). *The ethics of information*. Oxford: Oxford University Press.

Foucault, M. (1988). Technologies of the self. In L. H. Martin, H. Gutman, & P. Hutton (Eds.), *Technologies of the self: A seminar with Michel Foucault* (pp. 16–49). Amherst: University of Massachusetts Press.

Fuchs, C. (2014). *Social media: A critical introduction*. London: Sage.

Goffman, E. (1959). *The presentation of self in everyday life*. Garden City, NY: Doubleday.

Hursthouse, R. (1999). *On virtue ethics*. Oxford: Oxford University Press.

Johnson, D. (2001). *Computer ethics* (3rd ed.). Upper Saddle River, NJ: Prentice-Hall.

Lunt, P., & Livingstone, S. (2012). *Media regulation: Governance and the interests of citizens and consumers*. London: Sage.

Mackenzie, C. (2008). Relational autonomy, normative authority and perfectionism. *Journal of Social Philosophy, 39*(4), 512–533.

Markham, A., & Buchanan, E. (2012). Ethical decision-making and Internet research: Recommendations from the AoIR Ethics Working Committee (Version 2.0). Retrieved from http://www.aoir.org/reports/ethics2.pdf

Nissenbaum, H. (2010). *Privacy in context: Technology, policy, and the integrity of social life*. Palo Alto, CA: Stanford University Press.

Ong, W. (1988). *Orality and literacy: The technologizing of the word*. London: Routledge.

Plato. (1991). *The Republic of Plato*. (2nd ed.). (A. Bloom, Trans.). New York: Basic Books.

Rainie, L., & Wellman, B. (2012). *Networked: The new social operating system*. Cambridge, MA: MIT Press.

Rasmussen, T. (2014). *Personal media and everyday life: A networked lifeworld*. New York: Palgrave Macmillan.

Simmel, G. (1955). *Conflict and the web of group affiliations*. New York: The Free Press.

Simon, J., & Ess, C. (forthcoming). The ONLIFE initiative—A concept reengineering exercise. In E. Buchanan & M. Taddeo (Eds.), *Information societies, ethical enquiries*, special issue of *Philosophy & Technology*.

Stahl, B. (2006). Emancipation in cross-cultural IS research: The fine line between relativism and dictatorship of the intellectual. *Ethics and Information Technology, 8* (3), 97–108.

Vallor, S. (2010). Social networking technology and the virtues. *Ethics and Information Technology, 12*(2), 157–170.

Vallor, S. (2012). Flourishing on Facebook: Virtue friendship & new social media. *Ethics and Information Technology, 14*(3), 185–199.

Vallor, S. (2014). Moral deskilling and upskilling in a new machine age: Reflections on the ambiguous future of character. *Philosophy and Technology*, 1–18.

Westlund, A. (2009). Rethinking relational autonomy. *Hypatia, 24*(4), 26–49.

Wiener, N. (1954). *The human use of human beings: Cybernetics and society*. Garden City, NY: Doubleday Anchor. (Original work published 1950)

Wong, P. (2012). *Net recommendation: Prudential appraisals of digital media and the good life.* (Unpublished doctoral dissertation). University of Enschede, Enschede, The Netherlands.

CHAPTER THREE

Eudaimonia

Mobile Communication and Social Flourishing

RICH LING
NANYANG TECHNOLOGICAL UNIVERSITY, SINGAPORE

The ancient Greek concept of *eudaimonia* is translated as well-being, flourishing, or "the good life." It is the pursuit of satisfaction that goes beyond the hedonic physical pleasure of an individual. It is the use of self-discipline and prudence to achieve satisfaction that comes from practicing virtues and doing what is worth doing. Can this concept help us understand how mobile communication affects the lives of people and the structure of the societies they live in? Does the mobile phone help people flourish, or not? Does it bring broader moral issues into focus? Further, does this concept provide us, as researchers, a different lens to consider these issues?

To answer the first set of questions, there are undeniably positive effects of mobile communication for users around the world. It has been widely adopted and people feel that it makes a contribution to their lives. At the same time, the mobile phone exposes uncomfortable problems and issues. It causes stress; it is expensive; it jeopardizes our privacy; and it can threaten our sense of propriety. To be sure, the mobile phone is a two-edged sword. To answer the second question, eudemonia provides researchers with a useful perspective for examining the social consequences of mobile communication.

In this chapter, I draw examples around the globe to examine the adoption and use of mobile telephony. I begin with the case of Norwegian teens and how the mobile phone occasioned the reconsideration of their position in the society.

Then I discuss the effects of mobile telephony on women's role in Indonesia and China. Next, I shift to how the mobile telephony changes the sexual practices in Tanzania and Jamaica. Finally, I return to Norway and reflect on how the mobile phone is increasingly structured into the flux of society.

LESSONS FROM NORWEGIAN TEENS' ADOPTION OF MOBILE TELEPHONY

Teens' adoption of mobile telephony illuminates two classic issues: One on the moral questions of control and power, and the other on the consequences of ownership and use.

The Confrontation with Ideas of Propriety

When teens started to own and use the mobile phone, it was often seen as a moral breach (Nordal, 2000). Their use of the device contradicted social notions of teens' status and position in the society and, in a quiet Scandinavian way, it exacerbated opinion. In the mid-1990s, it was acceptable for people with legitimate needs to have a phone. However, teens' use violated many people's sense of what is appropriate. Interviews conducted in 1995 with Norwegian adults documented their concerns about the growing use of mobile phone among teens. One participant, Frank, noted,

> It is misused in a big way. For example, students have started to take it to school; they call during class. The fire alarm rings because of the calls. It is a normal occurrence. And people sit and talk to each other on the phone all the time. So just socially it is irritating. It is a good tool if it is used moderately and sensibly. Some think that it is dumb to use it in public and others think it is great. I often wonder why people talk so much on the phone. At stop lights they sit there and talk while they are waiting for a green light. I wonder what in the world they are talking about.

Teens are the definition of misusers for Frank. While including an urban legend in the narrative of teens' mobile phone adoption, Frank goes on to note the phone's potential for irritation. He recognizes its usefulness (if used properly), but he is nonetheless left in awe of potential for sociation enabled by the mobile phone both positively and negatively.

Following the theme of teens' use, other participants were dismissive. They felt that it was not appropriate for teens to have a mobile phone. It was expensive and teen users were simply seen as seeking social status. In sum, it was a rupture to their sense of decorum. Martin commented on the following situation,

> There was someone who told me that they had taken a bus after Christmas. There was a
> gang of boys that had been skiing and they sat there and talked [on a mobile phone] with
> their parents who were on their way to pick them up and told them that they were here and
> there. This is just awful.

Martin provides a telling statement since improved coordination—one of the most important contributions of mobile communication (Ling & Yttri, 2002)—is paired with the sense that it was "awful" to see teens using the phone. Other interviewees noted that the mobile phone was basically a status symbol for teens and that it was not necessary for them to own one.

Teens' use of the mobile phone was a type of Garfinklian breaching experiment (Woolgar, 2005) or a "Garfinkle machine."[1] In some ways, it has forced us to take stock of our sense of teens and their position in the society. There were good reasons for teens not to have one. There was the fear that they would run up unnecessary expenses and the fear that they would lose the device. There was the sense that providing teens with a mobile phone was giving them free rein and tantamount to losing insight into their activities and social lives.[2] The comments of the adults in the Norwegian focus groups point to how the mobile phone disrupted their sense of teens' position in society. In some cases, e.g., Frank's description of people talking while driving, time has shown that their comments were on point. However, when it came to teens' use as a marker of moral decline, the trajectory of practice has not followed their preliminary suggestions.

A Tool That Cuts in Two Directions

In the mid-1990s, it was rare for a teen to have a mobile phone in Norway or in many other countries for that matter. By 2005, however, analysis done by Statistics Norway was, quite literally, not able to find teens without a mobile phone by the middle of the decade in Norway (Vaage, 2008).[3] Clearly there had been a sea change. In less than a decade, the mobile phone had gone from being the foundation for moral panics (Thurlow, 2006) to being a matter of course. What had happened and what are the social consequences of this change? Moreover, did this change support the flourishing of mobile phone users?

Between 1995 and 2005, the mobile phone had found a secure place in quotidian life. The device was no longer only used by status seeking "first users." Rather, it had become embedded in the flux of daily life for both teens and adults. During this period, the price of phones had dropped as had the price of subscriptions. Also, the development of pre-paid subscriptions meant that teens could not run up unreasonably large bills.

Further, there had been a shift in attitudes toward mobile telephony. Parents recognized the advantages to being able to call their children whenever and wherever either interlocutor might be. Familial coordination was facilitated when, for example, teens could call parents when they finished soccer practice and tell them that they were going to a friend's house. This saved the parent a wasted trip. The fact that the mobile phone was available in the case of minor emergencies and that it allowed the marshalling of people as needed became obvious with use. These advantages weighed up against the earlier fears of losing control. In retrospect, it is clear that the efficiency in coordination and the continual access afforded by the device had trumped the fears voiced by adults in the mid-1990s. In addition, and perhaps most importantly, there had developed a reciprocal sense that one should be available to others and both teens and their parents would enjoy increased ability to plan if they both were available to one another. In the case of teens, this sometimes exposed power distinctions, such as the expectation that the teen needs to be available to parents whenever they might call, but the opposite was not necessarily always the case (Ling & Yttri, 2006).

To be sure, the mobile phone provided teens with their own channel of communication (Ling, 2008a). It allowed them to make their plans with friends for both innocent and not so innocent activities. On the one hand, it allowed teens to work out when and where to meet up for their sports and organized activities. On the other hand, research has shown that mobile communication in Norway covaried with sexual and other deviant activities (Ling, 2005; Pedersen & Samuelsen, 2003). Thus, while the mobile phone facilitated parent/child coordination when ferrying them to their different activities, it also made possible new forms of deviance such as bullying and chicanery (Raskauskas & Stoltz, 2007).

Teens' use of the mobile phone also meant that parents indeed did lose some contact with the social lives of their children. The shared landline phone in the home gave parents the opportunity for parents to have a quick chat with their children's friends when they called. That was washed away when the mobile phone became common. In addition, the mobile phone, and smart phones in particular, has exacerbated issues of cyberbullying, sexting, identity theft, being cheated out of money, etc. (Mascheroni & Ólafsson, 2014).

Thus, there was a tension between the positive and the negative side of the technology vis-à-vis teens. Is there evidence that introduction of the mobile phone led to human flourishing in the Global North? The device was certainly a tool that helped people to connect and facilitated their social activities in everyday life. A quick laugh or a bit of innocent gossip with our interlocutors, that is, in a Goffmanian sense, the mobile phone was used to forge social cohesion (Ling, 2008b). That said, the very same device also enabled various forms of deceit, decadence, and deviance. These tensions are pertinent in other parts of the world as well.

WOMEN AND MOBILE PHONES IN DEVELOPING COUNTRIES

Midwives in Indonesia

If we move the locus of the discussion to another corner of the globe, we find some of the same tensions, albeit framed around a somewhat different set of issues. It is instructive to look at the work of Arul Chib and Vivian Chen (2011) on mobile health (mHealth) in developing countries. They examined the effects of mobile phones given to rural midwives in Aceh, Indonesia, a religiously conservative and economically impoverished province that was severely affected by the 2004 tsunami.

Rural midwives were given mobile phones so that they could call doctors in the case of difficult births. The lower threshold for communication and greater immediacy was intended to enhance the provision of medical help. Indeed, this was one result of the trial. Chib and Chen (2011) reported that the lives of women and children were saved due to the mobile phone. Full stop. The mobile phone provided immediate and tangible benefits in the development of eudaimonia. The fact that it enabled more efficient delivery of healthcare confirms its role as a device that supports a better living situation for these people. It seems that this would cement the position of the mobile phone in the local society.

However, the local villages where these midwives operated had a gender hierarchy. Men have a relatively privileged position vis-à-vis that of women, regardless of the women's work or life situation. In addition, in the medical profession, male doctors generally hold positions of power when compared to midwives. Chib and Chen (2011) found that the mobile phone challenged the implicit understanding of hierarchical gender/professional relations. Their project illuminated dimensions of the power structure that had been tacitly in place up to that point. This was experienced along several dimensions. First, the mobile phone placed these midwives into a broader social structure beyond the local context, thus enhanced their social status.

Second, the mobile phone lowered the communication threshold between the midwife and the doctor. This meant that the doctor became, in theory, more accessible. Whereas the friction of distance and the inefficiencies of communication had limited this type of contact previous to the project, this was no longer the case. The doctors and the midwives were prompted to reconsider the appropriate form and frequency for interaction. For their part, the midwives quickly understood that their calls were not always welcomed by the doctors. Their experiences with the doctors could be confusing and sometimes characterized by the irritation of the doctors. The doctors also felt that the system allowed the midwives to cover up for one another and to mask their basic laziness. Thus, where the doctors had had the ability to literally distance themselves from the midwives' inquiries to the difficulty of access; this was no longer the case. They needed to draw on another register of control mechanisms (i.e., annoyance and derision) in order to rebalance the frequency of this interaction.

Third, the mobile phone also challenged the male-dominated family structure since it was often a woman who was the first family member to receive a mobile phone. Similar to the case with doctors, this arrangement challenged the tacit understandings of the woman's position in the social hierarchy. It exposed, and in some cases it inflamed, the existing system of power distribution. The mobile phone provided the women with an open communication channel that was not under the control of men. When seen from the perspective of the Muslim male, it perhaps violates his sense that he can protect his wife from an uncontrollable world. In a far more fundamental way than the Norwegian teens discussed above, giving the women a mobile phone was abdicating responsibility. According to Chib and Chen (2011), in some cases this was so acutely felt that the woman would solve the problem by simply giving the mobile phone to her husband or son.

Dagongmei in China

Some of the same tensions can be seen in Cara Wallis' (2011, 2013) work on young Chinese rural-to-urban migrant women *dagongmei*.[4] These women work in the low-level service sector. As with the midwives in Indonesia, the mobile phone serves as a two-edged sword. The mobile phone liberates them by allowing them to overcome spatial, temporal, physical, and structural boundaries. They are able to use it to (1) stay in contact with family back in their home villages, (2) play music and entertain themselves, (3) find jobs in the large urban centers, (4) develop and maintain *guanxi* or what Western scholars might call social capital, and (5) enact various forms of workplace resistance. Thus, the mobile phone is an empowering technology since it facilitates these *dagongmei's* pursuit of a career while also giving them access to extended social networks. Indeed, the mobile phone is an essential tool. It would be impossible for them to operate without one.

However, there is a Faustian dimension to having the device. Wallis (2011, 2013) has noted that employers use the mobile phone for surveillance and make demands on these *dagongmei* when they are not at work; they use the telephonic link to manipulate and control them. These *dagongmei* can be called to work at the whim of the employer and held accountable to their jobs even when they are not physically present. Wallis (2011) illustrated this with an episode where a mistrustful employer called a *dagongmei* to accuse her of theft when there had been a mistake in the storage of wares. Thus, while it is essential that the *dagongmei* have a mobile phone, it is not just a blessing. Despite the extended possibilities for these *dagongmei*, it is used to underscore the women's relatively powerless position in the society.

Mobile Phones and Transactional Sex

As noted for the Norwegian teens, the mobile phone is a device used for inter-actions between the genders. It has been used as a secret communication channel in the Philippines (Ellwood-Clayton, 2005) and in Palestine (Cohen, Lemish, & Schejter, 2008). In these situations, telephonic interaction between the young lovers was not sanctioned according to the traditional cultural norms. The mobile phone helped the couple to contact each other. At the same time, it evaded the culturally imbued control of parents and other authorities. Looking further in this direction, mobile telephony has allowed for new ways of organizing the more for-malized sex trade (Hearn, 2006) as well as the less formalized area of transactional sex (Hunter, 2002).

If we move our focus one more time to Dar es Salaam in Tanzania, the work of Laura Stark (2013) again underscores the potential and the problems with mobile communication. Stark (2013) examined the role of women in the Tandale neigh-borhood of Dar es Salaam. In that area, approximately 70% of the people live on less than one dollar a day. Stark describes how rather than engaging in an entre-preneurial economy where the goal was to develop a profit that was reinvested in the pursuit of further profit, she found a "buffering" economy where the goal was to develop a cushion that can be used when the next inevitable crisis arises. The crisis can take the form of sickness, loss of a job, theft, loss of a partner, or a host of other issues that are bound to arise.

In this chaotic situation the mobile phone facilitates the women's ability to gather information about different subsistence opportunities. It allows the women to organize small scale vending enterprises where, for example, they would go to the market and purchase a bucket of fish that they would fry and resell to people in the neighborhood for a small profit. As with the situation reported by Jagun et al. (2008) and entrepreneurship in Nigeria, the mobile phone allows the women stud-ied by Stark to quickly react to opportunities and to engage in a broader network of actors as they engage in these small scale entrepreneurial activities. It means that the network is more robust in the face of the cycles that buffer the lives of these women. Seen from the perspective of human flourishing, or at least survival, the mobile phone is a positive contribution in this context.

A less positive dimension, however, is that it facilitates what Hunter (2002) has called transactional sex. Unlike prostitution, transactional sex is not seen as a profession with relatively formalized roles, payments, etc. Rather, it is more dis-ordered with less distinct roles and more informal or bartered payment. An indi-vidual may have several different partners with whom they have occasional con-tact. The mobile phone facilitates transactional sex since it discreetly facilitates the management of the different relationships. The women in Tandale can maintain relationships with men who are not immediately proximate. The mobile phone is a channel through which they can keep the relationships secret from their spouses or parents.

As with the work by Stark, Heather Horst and Daniel Miller (2005, 2006) have found the same issue when studying mobile communication in Jamaica. Based on fieldwork among impoverished rural Jamaicans, the mobile phone has found its role in negotiating sexual interactions. They wrote:

> The maintenance of cross-sex relationships is probably the most visible use of cell phone technology in Jamaica. In practice the phone's use is symmetrical by gender, differing mainly in that men tend to see sexual relations as ends in themselves while women tend to associate them with other needs, most often economic survival. (Horst & Miller, 2005, p. 759)

Like the work of Stark in Tanzania, Horst and Miller describe how women use the mobile phone as a tool in a broader coping strategy where sex is exchanged for various commodities. In this context, Stark (2013) encapsulates the strategy by noting that the woman needs to provide food for her family; she uses sex to get that and her husband is happy that there is food when he comes home. Before the diffusion of the mobile phone, these transactions needed to be worked out face-to-face, which was a risk. The mobile phone provides the partners with an individualized communication channel that streamlines these more private "link up" (Horst & Miller, 2005) interactions and reduces the danger of being found out. So long as the interaction remains *sub rosa* the system can function. However, if there is a breach in the secrecy, perhaps occasioned by a rejected paramour, or if for some reason the woman becomes the subject of her husband's or eventually father's mistrust, the phone can provide telling evidence. Thus, while the phone is a tool used for economic survival, it also embodies a threat that can result in serious difficulties.

These vignettes from around the world underscore the relatively powerless situation of the women. The work of Stark (2013) and Horst and Miller (2005) also raise troubling questions regarding the stability of the family and the structure of monogamy for people living at the bottom of the economic pyramid. For example, do these findings suggest that stable monogamous family life is a luxury to be enjoyed only by people further up the economic ladder? It is clear that infidelity as well as alternative forms of family life exists in all societies, but it is troubling to think that systematized and distributed sexuality facilitated by mobile communication has become a feature of impoverishment.

Again, we come back to the suggestion that the mobile phone facilitates strategies that allow people to thrive, or at least to endure. The women in Indonesia, China, Tanzania, and Jamaica use the device as part of a strategy to endure and if they are lucky to thrive. They use it to do their jobs, to find work, and to provide for their families. The mobile phone also has an enormous potential for good. It can facilitate the provision of health care, inform people about issues related to their livelihood, and connect people to their nearest family and friends. At the same time, it can help institutionalize harmful practices, undercut trust in one another, and expedite the trafficking of what people see as socially destructive practices. In this process it can expose and inflame tacit systems of understanding and power thus making relatively powerless people vulnerable.

THE STRUCTURING OF MOBILE TELEPHONY INTO DAILY LIFE

One last stop on this world tour returns to Norway where I have done the majority of my work on the social consequences of mobile communication. In general, it is possible to see the positive edge of the sword, albeit with some problems.

The mobile phone has provided people in Scandinavia and in many other countries with an enhanced form of interpersonal interaction. As noted above, it has facilitated mundane coordination and helped in the everyday project of developing and maintaining social cohesion. The sum total of this is that the mobile phone has become structured into our expectations of one another. At the individual level, we get very concrete benefits from the device. It helps us to interact with others and to carry out different tasks; it has mapping functions to help us find our way and a camera to help us record a variety of events. We can keep track of our calendar and find the name of an important contact. In addition, there are thousands of apps that can be downloaded onto our smart phones that give us the ability to read the news, play Sudoku, update our Facebook page, listen to music, and watch a video. All of these things give us personal utility, that is, they help us to do things.

What sometimes gets lost in this welter of personal utility is that the device also connects us to our social networks. It is our link to others and their link to us. Looking beyond personal utility, there is reciprocal responsibility. We expect to be able to reach others just as they expect to be able to reach us. This web of mutual expectations is increasingly built into a generalized structuring of the mobile phone into the way that we approach daily life. That is, we are being slightly irresponsible to the generalized social sphere when we forget our phones at home or when we let the battery run out. We lose the personal utility of the device, but we also frustrate the efforts of others when they try to get in touch with us. Like Weber's iron cage (1930), there has developed a "soft coercion" to be available to one another given the flux of daily life.

In principle, we do not experience this as being an oppressive demand on one another. We accept the need to be available to one another just as we also understand when a friend or a family member is not able to take a call. However, there are situations where we exercise the need to be in touch. We were able to see this dimension to mobile communication in a horrific way during the bombing in Oslo on Friday, July 22, 2011. Our analysis of the mobile traffic data of 1.2 million users in the immediate aftermath of the bombing[5] showed that quite quickly after the bomb exploded people throughout the country used the mobile phone to check in with their strongest tie partner (Sundsøy, Bjelland, Canright, Engo-Monsen, & Ling, 2012). That is, when examining the data, within 30 minutes people all throughout Norway had checked in with the person that they had called the most during the previous three months. Almost immediately after that, they phoned the second most called person, and then third most. We traced this through the top five links and these links were called in approximately that order within the first minutes after the bomb had exploded.

We also saw an interesting geographical pattern with these calls (Sundsøy et al., 2012). As one would expect, there was an increase in calling within Oslo where people in the city could in fact hear the explosion. This was about four times the normal level of traffic for that time of day on a normal Friday. In addition, there was a similar increase in the number of calls into and out of the city. Again, this is not too surprising as people in Oslo were calling out to let people know that they were okay just as those outside the city were calling into Oslo to check on the status of their family and friends.

What is interesting, however, that that there was also a similar increase in people who did not live in Oslo calling to others who also lived nearby them. That is, people living in small cities outside of Oslo like Trogstad, Drammen, and Jessheim were calling to others living nearby. It is clear that these people did not immediately experience the bomb and there was no direct need to call others who were nearby to check on them. However, their calls show how the mobile phone is used in highly uncertain times to reach out to one another (Sundsøy et al., 2012). Interviews with people showed that these calls were made to alert one another about the situation and to share the news. Perhaps a neighbor had a relative in Oslo and so there was a vicarious "check in" taking place. While there were instrumental dimensions to the calls, there were also expressive dimensions. These were showing care and compassion to one another and in the process strengthening the web of social relations. They were people reaching out to each other in an uncertain moment. The mobile phone provided an immediacy in this situation that was not possible in other similar disaster situations (Erikson, 2012).

The bombing in Oslo was an exceptional social event that is mercifully rare. It affected a whole city and a whole country. At the same time, we also experience smaller scale dramatic events where we see many of the same mechanisms at work. When a child falls and breaks an arm, the immediate family needs to be mobilized to deal with the situation. Routines need to be interrupted and while one parent takes the child to the emergency room, the other needs to collect the other children from their activities and also needs to take over tasks that had been otherwise apportioned. Further, others need to be alerted to the situation and perhaps events need to be rescheduled (Ling & Donner, 2009). The mobile phone becomes an important tool in this work. Calls are made, people are contacted, and we work through the situation. The structure of our responses was laid bare in the case of the Oslo bombing. However, the same responses are obvious when looking at more localized emergencies.

In these situations, we also understand the degree to which the mobile phone is structured into the society. We expect to be able to reach one another when there is the need. It was an ominous portent if a child did not answer their phone during the events of July 22, 2011. It is clear that the mobile phone has changed the way that we understand and react to major social disasters (Erikson, 2012; Kavanaugh, Sheetz, & Kim, 2010; Vieweg, Palen, Liu, Hughes, & Sutton, 2008). If a family member is, for some reason, not available during a more localized emergency, it makes others' response to the situation more difficult. In all of this, there is a reciprocal expectation of availability that facilitates our ability to respond to the major (and the not so major) vagaries of daily life. In this way, we have structured the mobile phone into our daily expectations of one another. In this we gain the peace of mind that contributes to our well-being. That is, it helps us to thrive.

At the same time, there is the "soft coercion" of always being available. We need to have our phone with us. We need to pay for the subscription. We need to make sure the battery is charged, etc. This cost for people in Norway is generally not nearly as onerous as that of the midwives, domestic immigrants, or wives described above. The sword is not nearly as "two-edged" as in their case. There is still a sharp edge to it nonetheless.

SUMMARY

What exactly then is the mobile phone? It is a lens that exposes social structures; it is a change agent; it gives us individual utility; and it ties us into social structures in new ways. It is a cornucopia of information and it is a threat to our privacy. It exposes and enflames power structures just as it facilitates the organization of daily life. It has huge potential to help us at the individual level and at the social level. It can also fuel jealousy and mistrust and be seen as a threat to cultural systems that have been built up over many centuries.

Does the mobile phone provide for *eudaimonia*; for human and social flourishing? The answer seems to be yes and no. It is a tool that can help us care for one another but also a tool that can be used for nefarious ends. It can be used to make daily life easier for women in developing countries while at the same time it can be the locus of mistrust and misfortune. The mobile phone can be used to organize legitimate social protest (Jun, 2013) just as it can be to organize new types of crime and criminality (Agar, 2013).

Returning to the opening questions, what does this mean for the study of mobile communication? Can the notion of *eudaimonia* help us understand this phenomenon? In this case, the answer is simply yes. The notion of social flourishing gives us a framework with which to think about the social consequences of socio-technical phenomena. It suggests a research agenda and the eventual policy items that need to be considered that will sharpen the positive edge of the sword while dulling the negative edge.

NOTES

1. I am indebted to Charles Ess for this idea.
2. This is not a new idea, indeed the Lynds (1929; see also Brandon, 1981; Grinter & Palen, 2002) reported the same issue associated with the landline phone in their 1920 study of Middletown.
3. Statistics Norway is Norway's national statistical bureau. Thus, this is not a casual statistic done by a sensation seeking fly-by-night operation. It is the results of the best statistical sampling organization in Norway with decades of experience and the resources with which to do many call-backs to the people in their sampling frame. The fact of the matter was that they were not able to find teens between ages of 15 and 20 without a mobile phone.
4. Meaning "working little sister" or "maiden worker."
5. We examined only the events in Oslo and not the shootings on Utøya. This was because there was a serious potential to breach privacy in the latter event where the bombing in Oslo did not have the same problem. Further, in broad terms, the event on Utøya was much more limited in a telephonic sense than was the bombing. In that case there were several hundred people directly involved, where the bombing in Oslo involved or was obvious to hundreds of thousands of people.

REFERENCES

Agar, J. (2013). *Constant touch: A global history of the mobile phone*. London: Icon Books.
Brandon, B. B. (1981). *The effects of the demographics of individual households on their telephone usage*. Cambridge, MA: Ballinger.
Chib, A., & Chen, V. (2011). Midwives with mobiles: A dialectical perspective on gender arising from technology introduction in rural Indonesia. *New Media & Society, 13*(3), 486–501.
Cohen, A., Lemish, D., & Schejter, A. M. (2008). *The wonder phone in the land of miracles: Mobile telephony in Israel*. Cresskill, NJ: Hampton Press.
Ellwood-Clayton, B. (2005). Desire and loathing in the cyber Philippines. In R. Harper, L. Palen, & A. Taylor (Eds.), *The inside text: Social, cultural and design perspectives on SMS* (pp. 195–222). London: Klewer.
Erikson, K. T. (2012). *Everything in its path*. New York: Simon & Schuster.
Grinter, R., & Palen, L. (2002). Instant messaging in teen life. In *Proceedings of the ACM Conference on Computer Supported Cooperative Work (CSCW 2002)*. New Orleans, LA: ACM.
Hearn, J. (2006). The implications of information and communication technologies for sexualities and sexualised violences: Contradictions of sexual citizenships. *Political Geography, 25*(8), 944–963.

Horst, H. A., & Miller, D. (2005). From kinship to link-up: Cell phones and social networking in Jamaica. *Current Anthropology, 46*(5), 755–778.

Horst, H. A., & Miller, D. (2006). *The cell phone: An anthropology of communication.* Oxford: Berg.

Hunter, M. (2002). The materiality of everyday sex: Thinking beyond "prostitution." *African Studies, 61*(1), 99–120.

Jagun, A., Heeks, R., & Whalley, J. (2008). The impact of mobile telephony on developing country micro-enterprise: A Nigerian case study. *Information Technologies and International Development, 4*(4), 47–65.

Jun, L. (2013). *Mobilized by mobile media: How Chinese people use mobile phones to change politics and democracy.* Copenhagen University, Copenhagen.

Kavanaugh, A.L., Sheetz, S., & Kim, J. B. (2010). Cell phone use with social ties during crises: The case of the Virginia Tech tragedy. In *7th International Conference on Information Systems for Crisis Response and Management (ISCRAM).* Seattle, WA.

Ling, R. (2005). Mobile communications vis-à-vis teen emancipation, peer group integration and deviance. In R. Harper, L. Palen, & A. Taylor (Eds.), *The inside text: Social, cultural and design perspectives on SMS* (pp. 175–194). London: Klewer.

Ling, R. (2008a). Mobile communication and teen emancipation. In G. Goggin & L. Hjorth (Eds.), *Mobile technologies: From telecommunications to media* (pp. 50–61). New York: Routledge.

Ling, R. (2008b). *New tech, new ties: How mobile communication is reshaping social cohesion.* Cambridge, MA: MIT Press.

Ling, R., & Donner, J. (2009). *Mobile communication.* London: Polity.

Ling, R., & Yttri, B. (2002). Hyper-coordination via mobile phones in Norway. In J. E. Katz & M. Aakhus (Eds.), *Perpetual contact: Mobile communication, private talk, public performance* (pp. 139–169). Cambridge: Cambridge University Press.

Ling, R., & Yttri, B. (2006). Control, emancipation and status: The mobile telephone in teens' parental and peer relationships. In R. Kraut, M. Brynin, & S. Kiesler (Eds.), *Computers, phones and the Internet: Domesticating information technology* (pp. 219–234). Oxford: Oxford University Press.

Lynd, R. S., & Lynd, H. M. (1929). *Middletown: A Study in modern American culture.* New York: Harcourt Brace.

Mascheroni, G., & Ólafsson, K. (2014). *Net children go mobile: Risks and opportunities* (2nd ed.). Milano: Educatt.

Nordal, K. (2000). *Takt og tone med mobiltelefon: Et kvalitativt studie om folks brug og opfattelrer af mobiltelefoner.* Universitetet i Oslo, Institutt for sosiologi og samfunnsgeografi.

Pedersen, W., & Samuelsen, S. O. (2003). Nye mønstre av seksualatferd blant ungdom. *Tidsskrift for Den Norske Lægeforeningen, 21*(6), 3006–3009.

Raskauskas, J., & Stoltz, A. D. (2007). Involvement in traditional and electronic bullying among adolescents. *Developmental Psychology, 43*(3), 564–575.

Stark, L. (2013). Transactional sex and mobile phones in a Tanzanian slum. *Suomen Antropologi: Journal of the Finnish Anthropological Society, 38*(1), 12.

Sundsøy, P. R., Bjelland, J., Canright, G., Engo-Monsen, K., & Ling, R. (2012). The Activation of Core Social Networks in the Wake of the 22 July Oslo Bombing. In *2012 IEEE/ACM International Conference on Advances in Social Networks Analysis and Mining (ASONAM)* (pp. 586–590). doi:10.1109/ASONAM.2012.99

Thurlow, C. (2006). From statistical panic to moral panic: The metadiscursive construction and popular exaggeration of new media language in the print media. *Journal of Computer-Mediated Communication, 11*(3), 667–701.

Vaage, O. (2008). *Mediabruks undersøkelse, 2007*. Oslo: Statistics Norway.

Vieweg, S., Palen, L., Liu, S. B., Hughes, A. L., & Sutton, J. N. (2008). *Collective intelligence in disaster: Examination of the phenomenon in the aftermath of the 2007 Virginia tech shooting.* In *5th International Conference on Information Systems for Crisis Response and Management (ISCRAM)*. University of Colorado.

Wallis, C. (2011). Mobile phones without guarantees: The promises of technology and the contingencies of culture. *New Media & Society, 13*(3), 471–485. doi:10. 1177/1461444810393904

Wallis, C. (2013). *Technomobility in China: Young migrant women and mobile phones*. New York: NYU Press.

Weber, M. (1930). *The Protestant ethic and the spirit of capitalism*. London: Routledge.

Woolgar, S. (2005). Mobile back to front: Uncertainty and danger in the theory-technology relation. In R. Ling & P. E. Pedersen (Eds.), *Mobile Communications: Re-negotiation of the Social Sphere* (pp. 23–44). London: Springer-Verlag.

CHAPTER FOUR

Meaningfulness and Entertainment

Fiction and Reality in the Land of Evolving Technologies

MARY BETH OLIVER
PENNSYLVANIA STATE UNIVERSITY, USA

JULIA K. WOOLLEY
CALIFORNIA POLYTECHNIC STATE UNIVERSITY,
SAN LUIS OBISPO, USA

Foundational research in entertainment psychology has tended to focus its attention on fictional, narrative content. With some notable exceptions such as studies of sporting events, most entertainment scholarship has examined various types of stories, including comedy, mystery, horror, and tragedy, among many others (for overviews, see Klimmt & Vorderer, 2009; Oliver, 2009; Zillmann & Vorderer, 2000). Understandably, this research has also tended to examine entertainment appearing in more traditional media, including television, literature, and film. Finally, entertainment scholarship has been heavily influenced by theories making assumptions about individuals' selection and enjoyment of content, placing an emphasis on hedonic considerations and on the importance of beloved characters being portrayed as triumphant or antagonists as defeated (Raney, 2006; Zillmann, 2000).

With this background in mind, more recent scholarship has begun to broaden the conceptualization of entertainment, noting that in addition to giving rise to pleasures akin to hedonic gratifications, it may also provide individuals with meaningful and often deeply enriching experiences, akin to eudaimonic experiences (Oliver & Raney, 2014). This growing scholarship has employed a variety of perspectives to examine broadened notions of entertainment gratifications and processes. However, with a few exceptions discussed below, studies in this growing area have generally continued to employ fictional narratives in traditional mediums.

Though television and film may still constitute a large segment of the entertainment landscape, newer and emerging technologies are undoubtedly playing increasingly important roles in our entertainment lives. These technological changes not only allow audiences greater access to entertainment content, but also introduce a host of new variables that may not have been central to our prior theorizing. For example, concepts such as *presence* (Lee, 2004) and *interactivity* (Klimmt, Vorderer, & Ritterfeld, 2007) are now commonly discussed by media psychologists but were rarely heard of in relation to traditional media such as film or television.

The myriad ways that technologies have changed the entertainment experience are impossible to detail in one brief chapter. Our goal is much more modest: to continue the recent discussion of "meaningful" media, but with a focus on newer entertainment technologies. Specifically, our over-arching question concerns the extent to which technological changes in the media landscape hinder or facilitate eudaimonic entertainment experiences. In considering this question, we opt to broaden our conceptualization of entertainment so as to include entertainment afforded by newer media platforms (Reinecke, Vorderer, & Knop, 2014). In particular, our discussion recognizes that fictional narrative is only one small slice of media entertainment, with increasingly larger segments being populated by mobile communication, social media, and user-generated content. As a result, media entertainment is no longer only (or perhaps even *primarily*) about fiction, but is instead more heavily situated in "real life"—entertainment that we take with us in our day-to-day lives, entertainment that is created from snapshots of our real-life moments, and entertainment that is focused on "real" people and their "real" stories.

This chapter considers how newer technologies are increasingly blurring entertainment and reality, and what this blurring may imply about meaningful entertainment experiences and meaningful "real life" experiences. We begin with a brief discussion of the concept of reality, how it has been studied in extant media research, and why it may be important for research on meaningful media in particular. We then turn to the ways that newer technologies may undermine our search for eudaimonic fulfillment, and finally, how they may present new and unique opportunities to experience heightened levels of insight, inspiration, and appreciation.

REALITY, MEDIA ENTERTAINMENT, AND MEANING

The relationship between media entertainment and "real life" has been the focus of a diversity of research from a variety of vantage points. For example, the importance of reality in media scholarship has grown tremendously over the last several decades, as "reality" genres have become a staple of the television line-up. Whereas early reality programming focused on crime-related topics, this genre now addresses almost any topic imaginable, ranging from dating, to home improvement, to beauty pageants. Scholarship in this area has examined not only how "reality" portrayals compare to actual reality (Oliver, 1994), but also the uses and gratifications that viewers have for consuming this often maligned but nevertheless very popular form of entertainment (Nabi, Biely, Morgan, & Stitt, 2003).

Additional entertainment scholarship has focused on the tendency for individuals to confuse fantasy and reality in media portrayals, and the implications of these perceptions for viewers' emotional responses and engagement. For example, early anecdotal accounts of motion picture audiences described viewers screaming and running from the cinematic scene of an oncoming train in the Paris showing of *L'Arrivée d'un Train en Gare de la Ciotat* in 1895 (Tan, 1996). Although such confusion may seem quaint if not silly, contemporary audiences in 3D and IMAX theaters exhibit similar responses, with viewers ducking as images of bullets whiz past their heads and succumbing to motion sickness when floating about in zero-gravity outer space (Marshall, 2013). Of course, adult audiences generally recognize the fictionality of many entertainment offerings (despite their body's confusion), but media scholarship suggests that such differentiation is something that develops over time—children, in particular, seem prone to confusing fiction with reality, with such confusion resulting in specific and unique emotional responses among child versus adult viewers (Cantor, 2009).

The importance of perceived realism in entertainment experiences is perhaps most familiar via the use of the phrase "suspension of disbelief." In short, this phrase refers to the idea that although entertainment consumers understand the fictionality of most narratives, they willingly forgo their critical stance so as to engage with and enjoy stories. In contrast, in overviewing and critiquing this assumption, Busselle and Bilandzic (2008) argued that rather than effortfully abandoning scrutiny of the reality of texts, individuals typically engage narratives with the assumption of "realism" in place. That is, individuals tend to accept the reality of the narrative world and to not question or critique it unless confronted with inconsistencies between the narrative and external reality, or with inconsistencies in the narrative itself.

Together, extant research on media and "real life" draws attention to the importance that perceived realism plays in viewers' cognitive and emotional responses to entertainment, though the concept of realism has yet to play a central role in more recent work on entertainment pertaining to meaningfulness and well-being. However, we believe that this growing area of entertainment scholarship could benefit from a greater recognition of the notion of reality (and associated concepts such as realism, factuality, fictionality, etc.) in several ways. First, we note that media entertainment, in addition to depicting real life, can become *part* of real life—intertwined into how we spend our days, capturing and transmitting life moments, and reflecting and affecting social relationships. Second, we believe that incorporating the notion of realism or the "real" may position scholars to explore additional and nuanced ways that media may play a role in feelings of meaningfulness and well-being. In particular, although the notion of "realism" can imply many things, including accuracy, representativeness, or precision, it can also connote genuineness, truth, and sincerity. These latter concepts suggest the value we place on experiences, people, and things we understand to be actual, basic, and fundamental. We are moved by authenticity, we seek insight into truth, and we find heightened connectedness to humanity through our sincere sharing of basic virtues. Hence, meaningfulness and "real life" appear to intersect in non-trivial ways that may help us to broaden our understanding of both hedonic and eudaimonic entertainment experiences.

With these additional notions of media and "realism" in mind, we now turn to the ways that emerging entertainment technologies may intersect with "real life," and what these intersections imply for individuals' sense of well-being and their experience of meaningfulness. We first consider the potential negative effects of these technologies on such outcomes, and we end our chapter by exploring possible avenues by which they may enrich our lives.

NEGATIVE SIDES OF EVOLVING ENTERTAINMENT TECHNOLOGIES

The possible harmful aspects of evolving media technologies are voluminous and diverse. Hours spent playing video games rather than running about the playground are thought to be one contributor to childhood obesity (Vandewater, Shim, & Caplovitz, 2004), the demands of multi-tasking are argued to have hampered our ability to sustain focused attention (Ophir, Nass, & Wagner, 2009), and the anonymity of online postings are said to be one contributing factor to the ubiquity of flaming (Siegel, Dubrovsky, Kiesler, & McGuire, 1986). While acknowledging the multitude of ways that evolving entertainment technologies may harm our sense of well-being, we focus our attention on two potential outcomes that reflect the intersection of entertainment with "real life": the intrusion of mobile technologies into real-life interactions, and the blurring (and therefore trivialization) of meaningful life moments and media entertainment.

Diminishment of Real-Life Interactions

The mobility of media technologies means that we are rarely "free" from our entertaining diversions. Recent statistics illustrate that we are tethered to our communication devices, reaching for our mobile phones approximately 150 times a day

(Meeker & Wu, 2013). We sleep with our smart phones by the bed stand, we compulsively check them even if there is no incoming message, and we feel lost if we leave home without our mobile sources of contact (Pew Research Internet Project, 2014; Vorderer & Kohring, 2013). Although the use of mobile media in and of itself undoubtedly presents a host of gratifications, the constant stream of media communication runs the risk of directing our attention away from actual, personal, face-to-face interactions that form the basis of relational development, social support, and ultimately, meaningful encounters. As Turkle (2011) noted:

> Media technology has made each of us "pauseable." Our face-to-face conversations are routinely interrupted by incoming calls and text messages. …In the new etiquette, turning away from those in front of you to answer a mobile phone or respond to a text has become close to the norm (p. 161).

Despite the idea that it is now routine to have interpersonal communication interrupted by mobile technologies, people seem to continue to find this behavior disruptive and annoying. A recent national U.S. poll regarding perceptions of etiquette reported that sending emails and texting during meals was the #1 annoyance, with 56% of respondents identifying it as the activity that they would most like others to stop doing (60 Minutes/Vanity Fair Poll: Etiquette, 2014). Additional experimental research provides further evidence for the deleterious effects of mobile media on face-to-face interaction. Przybylski and Weinstein (2013) asked unacquainted individuals to hold a 10-minute conversation in the presence or absence of a mobile phone that was placed on a nearby table. The mere presence of the phone resulted in lower levels of trust, partner empathy, and perceived relational quality, but particularly when the participants were instructed to discuss meaningful topics rather than a casual topic. Misra, Cheng, Genevie, and Yuan (2014) recently replicated these results in naturalistic settings (e.g., coffee shops) by recording whether or not either participant placed a cell phone on the table during their conversation, along with the participants' perceptions of the quality of the conversational interaction. Although the authors were careful to note that their findings do not provide an unambiguous explanation for why the presence of mobile phones appears to detract from the perceived quality of interpersonal communication, they speculated that the distraction of mobile phones may diminish one's ability to detect subtle but meaningful conversational cues such as eye contact and facial expression. Placing one's mobile phone on the table may also signal to one's conversational partner that their conversation is secondary in importance to other concerns, and/or that the intimacy of their conversation is subject to disruption or intrusion.

Additional research clearly needs to be conducted to further explore how and why mobile communication devices appear to diminish the meaningfulness of face-to-face social interactions. But in the meantime, it appears that individuals have an intuitive sense that foregoing mobile devices is healthy for relationships. For example, "phone stacking" is an increasingly popular activity in which people at a social gathering (often a meal) agree to rid themselves of their mobile devices by placing them in a designated place (e.g., stacked in the middle of the table) so as to enhance social interaction. The rules of the game typically designate some type of "punishment" (e.g., paying for the meal) to the first person who touches her/his phone (Phone stack, 2014). As one blogger noted, "It used to be cool to have a cell phone, to always be able to reach out and be reached. Now it's disconnecting that gives us satisfaction.... I want to disconnect so I can reconnect" (Sweet, 2013).

Blurring and Trivialization of Meaningful Life Moments

Our discussion thus far has noted how life is impacted when people direct their attention *away* from real life and *toward* technological diversions. In addition to these examples, though, we believe that there is another way that technologies impact real life—through the *blurring* of mediated and real-life experiences. Insofar as technologies are frequently used for purposes of entertainment frequently consumed for pleasure or diversion, it is our concern that this blurring ultimately serves to trivialize real-life experiences that may otherwise be deeply meaningful to individuals (Raney & Oliver, 2014).

The idea that actual, non-mediated experiences may be affected by dominant communication technologies is an observation expressed by Neil Postman (1986) in his discussion of television:

> Television is our culture's principal mode of knowing about itself. Therefore—and this is the critical point—how television stages the world becomes the model for how the world is properly to be staged. It is not merely that on the television screen entertainment is the metaphor for all discourse. It is that off the screen the same metaphor prevails. (p. 92).

For Postman, the shift from print to television meant that aspects of our actual lives—politics, news, education, and religion—were transformed so as to be consistent with the primary means of communication.

In the age of social media and mobile entertainment, technologies continue to transform "actual" lives, as our actual lives are frequently what is photographed, tweeted, recorded, and disclosed. Selfies are now a commonplace phenomenon, party and vacation photos are routinely posted online, and videos of family and friends are abundantly shared (Hollenbaugh & Ferris, 2014; Mendelson & Papacharissi, 2011). What was once intimate and potentially meaningful is now disclosed to thousands if not millions instantaneously for purposes of amusement, if not ridicule. Marriage proposals, including those that were declined, become the fodder for public viewing and commentary, and even childbirth is recorded and then posted and tweeted for the viewing pleasure of the online audience.

Yet the blurring of private and public spaces does not just mean that real life is now disclosed more widely. Rather, we argue (as did Postman) that entertainment technologies are now transforming real life, including, at times, its most meaningful moments, so that real life "looks the part" for its posting on Facebook and its sharing on Tumblr. The YouTube video *I Lost my Phone* (written by and starring Charlene deGuzman, 2013) illustrates these types of changes—friends on a beach focus their attention on a phone for a selfie rather than on the shoreline, guests at a birthday gathering turn into a flock of paparazzi as they record video of the blowing out of candles. Although these events—a visit to the ocean or a birthday party—would have likely happened in the absence of technology, the way the participants are posed, the way the lighting is "just right" for the camera, and the way that everyone is sure to smile at exactly the perfect time all suggest that these real life "events" are somehow transformed and staged so as to appeal to the hypothetical viewers/users who will ultimately become the entertainment audience when these moments are shared with the world.

Further, we believe that, at times, these types of "real life" events may be *created* with the idea of entertaining social media in mind. In other words, pranks are played, songs are sung, and friends and family are posed in ways that might have never been played, sung, or posed in the not-too-distant past, but are done so now in the age of social media because the events "fit" the emerging form of entertainment. As a consequence, the concept of a "pseudo event" needs to be broadened to encompass not only the political staging of events, but also the everyday staging of events that make for good media and good entertainment—a sort of "pseudo life event."

Finally, the constant opportunity for everyday life to become entertainment content suggests that the world may now frequently be viewed through the lens of this entertainment framework, even in instances where people are not specifically using entertainment technologies. So now, a beautiful rainbow is seen through the lens of a potential posting in social media, a baby's first tooth is recognized as a "perfect" posting on YouTube, and even something as incomprehensibly tragic as the collapse of the Twin Towers on 9/11 is perceived as looking "just like a movie." In other words, meaningful life moments are seen at "meta level"—as how they *would* look were they posted or shared, or how they *will* be perceived and appreciated by one's social network. As a result, rather than living *in the moment* during these meaningful encounters, evolving entertainment technologies are teaching us how to live in hypothetical worlds in which the electronic representation of events is somehow more important than actual events themselves.

Clearly future research is needed to evaluate our argument that the ever-present opportunity to photograph or record life events detracts from their meaningfulness. However, recent scholarship suggests that, at a minimum, the digital recording of our lives may ultimately serve to diminish our memory of life events.

Henkel (2014) found that when participants were instructed to take photographs of whole art works in a museum rather than simply looking at the objects, memory of the objects themselves and of their detail was diminished. Henkel interpreted these findings as suggesting that taking photos of whole objects rather than focusing attention and carefully observing the objects results in people relying on photographs, rather than on their own observations and memories, to "remember for them" (p. 401). Applied to emerging entertainment technologies, these results imply that although our constant recording of lived moments may represent a thorough archiving, it may ultimately detract from our careful observation, memory, and hence experience of life events that might otherwise be deeply meaningful. Much like the "phone stacking" game, there seems to be growing public awareness of this problem as well. For example, the "unplugged wedding," a phenomenon where couples ask their invitees to refrain from taking photos or using social media during their event, has become increasingly popular. In an article about this phenomenon from the Huffington Post, one photographer laments the omnipresence of attendees holding up phones, cameras, even large iPads up to capture key moments in the wedding ritual (Corey Ann, 2013). Although this particular article clearly reflects a practical concern on the part of the photographer regarding how these intrusions ruin professional photographs, many couples also cite the issue of guest "presence" in their decision to go "unplugged." By conducting a preliminary search of Google images for "unplugged wedding," one can find many appeals to guests inscribed on invitations or chalkboards (e.g., "too often we get distracted by all of technology's new toys," "we would love for you to be completely immersed in the ceremony with us," and "we want our family and friends present during this moment of our lives"). While concerns for expensive photography are likely at play, we believe this trend also reflects a growing concern with protecting meaningful life moments from the intrusions of emerging entertainment technologies.

Summary

To summarize, changes in entertainment technologies have resulted in the omnipresence of entertainment during our day-to-day interactions, as well as the potential transformation of our interactions into entertaining content. These changes represent challenges to our realization of meaningfulness by distracting us from meaningful encounters, and by encouraging us to see and create life events through the lens of entertaining media. Ultimately we think this state of affairs threatens to devalue what we normally hold dear. At a minimum, we believe that as media scholars and as consumers of media, we need to be vigilant of these potential harms. At the same time, though, we also believe that it is important for us to consider the new opportunities that emerging technologies may afford in our search for meaning. The next section considers the brighter sides of newer forms of entertainment, and what they may imply about our quest for meaningfulness.

POSITIVE SIDES OF EVOLVING ENTERTAINMENT TECHNOLOGIES

Although entertainment is typically understood in terms of the hedonic pleasures it provides, people appear to deeply value forms of entertainment that encourage deeper insight and contemplation, even when such contemplation may be emotionally draining or even distressing. The findings from a recent archival study of over 500 film titles spanning three decades is consistent with this argument—movies that were action-packed and exciting were most successful in terms of box-office revenues, but ones that were dramatic, contemplative, and emotional elicited the greatest popular and critical acclaim (Oliver, Ash, Woolley, Shade, & Kim, 2014). In addition to encouraging greater insight and reflection, meaningful entertainment also appears to encourage us to be outward looking—to see others in a new light and to be inspired by the human spirit. We are humbled by the courage and conviction of Morgan Freeman's portrayal of Nelson Mandela in the movie *Invictus*, and we are moved by the bravery of Roberto Benigni's character in *Life Is Beautiful*.

The idea that entertainment can provide us with insight and inspiration—even when such experiences may be associated with mixed or negative affect—is central to our understanding of audiences' appreciation of eudaimonic entertainment experiences. In keeping with the theme of "realism" that forms the basis of this chapter, in this section we consider how more "realistic" experiences of moral emotions afforded by interactivity and by user-generated content may hold potential for heightening the meaningfulness of entertainment experiences.

Interactivity, Morality, and Heightened Appreciation

In discussing the nature of meaningful entertainment, Oliver and Bartsch (2011) argued that portrayals of moral considerations may be particularly important. Drawing on Aristotelian notions of eudaimonia (Aristotle, trans. 1931), these authors stated:

> [E]ntertainment can be understood as increasingly meaningful when it focuses to a greater extent on questions of human moral virtues, it demonstrates such virtues (or the ramifications of the lack thereof), it teaches or inspires insight into these virtues, or it causes the viewer to contemplate them and what it means to live a "just" or "true" life (p. 31).

Additional scholarship on media entertainment continues to stress the importance of moral considerations in affecting audiences' processing and responses. Most recently, Tamborini developed the MIME (Model of Intuitive Morality and Exemplars) that argues for the importance of moral depictions in audiences' reactions (Tamborini, 2011, 2013). Specifically, Tamborini argues that when entertainment features moral portrayals that are straightforward and that uphold one's moral considerations, such portrayals are easy for audiences to process and result in heightened enjoyment. In contrast, entertainment featuring more complex moral portrayals or depictions of conflicting moral considerations results in heightened deliberation and with greater appreciation of the entertainment provided that moral concerns are minimally acceptable to the viewer (Lewis, Tamborini, & Weber, 2014).

With this background in mind, what might evolving entertainment technologies imply about audiences' experiences of meaningfulness arising from moral concerns? We believe that one possible outcome is that the heightened interactivity associated with newer forms of entertainment and the "realism" of the associated content may make moral considerations more salient and may afford different types of emotional experiences, including ones that may be perceived as particularly meaningful. Below we describe two ways that interactivity may contribute to this outcome.

First, heightened interactivity means that rather than simply *observing* the moral conflicts that may be part of a storyline, the user/viewer/gamer of newer forms of entertainment may be in a position to *make* difficult moral decisions in order for the storyline to progress. For example, in the videogame *Heavy Rain*, the primary protagonist played by the gamer is forced to make difficult decisions that have ramifications throughout the rest of the narrative. By drawing attention to the potential importance of interactivity in the experience of moral conflict, we do not mean to suggest that it is unimportant in non-interactive entertainment. Indeed, examples abound of protagonists (and antagonists) facing difficult moral decisions: Atticus Finch (*To Kill a Mockingbird*) decides to stand up for social justice in the face of racism, and Walter White (*Breaking Bad*) makes the life-changing decision to make methamphetamine. Yet videogames and other forms of interactive entertainment may present a particularly salient opportunity for audiences to experience moral conflict, as the narrative can depend on the user/gamer having to deliberate about moral considerations as part of the gaming experience. As a result, we believe that it is potentially profitable to examine the idea that interactive entertainment has the opportunity to be particularly meaningful via its realistic and salient presentation of questions of morality.

The second and related way that we believe that the interactivity in emerging forms of entertainment may affect meaningful experiences is via the broadened *types* of emotion that may be possible. Whereas watching characters display admired values or (in contrast) demonstrate a lack of moral virtue can elicit deep emotions such as admiration, contempt, and elevation (which we discuss more below), actually *making* the moral decisions *as the character* may arouse feelings that require greater agency. For example, the experience of *guilt* implies that one may have been *responsible* for a bad decision that hurt others, the experience of *regret* implies wishing that one had *not made* a bad past decision, and the feeling of *pride* implies that one embraces the *responsibility* of a victory or success. Although all of these emotions may be experienced vicariously by viewing characters and the decisions that they make, we suggest that engagement *as the character* rather than as only the viewer may qualitatively change the nature of the affective or emotional experience. For example, seeing a character perform a heinous act may elicit feelings of disgust or disdain, but performing the heinous act oneself may elicit feelings of guilt or shame (Hartmann, Toz, & Brandon, 2010). Recent research suggests that these experiences of guilt can, in turn, make players more morally sensitive, increasing the salience of moral considerations regarding care and fairness (Grizzard, Tamborini, Lewis, Wang, & Prabhu, 2014). To the extent that interactive media effectively and realistically result in individuals feeling responsible for their actions and the outcomes that follow, we suggest that these newer forms of entertainment may present a wider array of opportunity by which entertainment can elicit insight and deeper understanding into the human condition and all its experiences, including the beautiful ones (such as pride) and the difficult ones (such as regret).

Elevation and "Real" Images of Moral Beauty

The importance of moral considerations to feelings of meaningfulness is particularly relevant in terms of the emotion of *elevation*. Haidt and his colleagues described elevation as the affective response from witnessing others who display or embody acts of moral beauty such as "charity, gratitude, fidelity, [and] generosity," with such responses also giving rise to the desire to emulate or exemplify virtuous or noble traits (Algoe & Haidt, 2009, p. 106; see also Haidt, 2003; Schnall, Roper, & Fessler, 2010). Although numerous studies of elevation have employed media texts (e.g., newspaper stories, television clips) as experimental stimuli (e.g., Freeman, Aquino, & McFerran, 2009; Silvers & Haidt, 2008), only recently have studies begun to specifically focus on *media* as a source of elevation (Oliver, Ash, & Woolley, 2013). For example, Oliver, Hartmann, and Woolley (2012) found that movies featuring portrayals of moral values were particularly likely to be named as "meaningful" movie experiences, with such portrayals associated with heightened levels of elevation that were characterized in terms of mixed affect, physiological sensations (e.g., chills, tears), and motivation to engage in benevolent behaviors. Likewise, the aforementioned archival study of popular and meaningful films (Oliver et al., 2014) points to the idea that many meaningful movies may be elevating to audiences, as critical acclaim was higher for films described by terms such as *touching, emotional, uplifting, hopes*, and *about the human spirit*.

These recent studies suggest that entertainment media are capable of eliciting elevation, and that such experiences are valued by audience members. But what do evolving technologies imply about the ability of newer forms of entertainment to inspire and move audiences? To our knowledge, research has yet to directly compare the effects of entertainment formats on this specific audience reaction. However, recent research that employed user-generated content (YouTube videos) found that these newer and popular videos were not only capable of eliciting elevating affect, but also of leading to heightened feelings of optimism about and connectedness to humanity per se (Oliver et al., in press).

We recognize that entertainment on newer media platforms such as YouTube may frequently show the trivial sides of life (e.g., "Epic Fails") or, worse yet, the antagonistic or even hostile ways that humans may treat one another. At the same time, though, we also hold out hope that evolving technologies may have promise of being particularly meaningful and elevating, partially due to the "realism" of their depictions. We have never-before-seen opportunities to witness and create media experiences that encourage us to see the beautiful sides of the human spirit among real, actual people—including among those who may be geographically distant. Inspiring media content abounds on sites aimed at enhancing social good (e.g., UpWorthy, http://www.upworthy.com/) and is very salient in user-generated sites and on social media platforms where it quickly goes viral. Kid President encourages us in his pep talks (http://youtu.be/l-gQLqv9f4o), musicians around the world elevate us by playing together for issues of social change (http://youtu.be/Us-TVg40ExM), and a video of an elderly lady finding her beloved dog in the debris left by a tornado moves us in its depiction of simple and pure love (http://youtu.be/BUc1ZgXiQGs). What we believe may be consequential is that many of the inspiring depictions found in newer entertainment platforms are of real people, real displays of kindness and generosity, and real acts of patience and acceptance. This connection with real life that we bemoaned previously in terms of its potential negative outcomes may conversely serve a noble purpose in its authenticity—heightening our favorable perceptions of fellow human beings, reminding us of our shared values and our celebrated virtues, and consequently paving the way to greater connectedness and compassion. Though our argument about the importance of reality is speculative at this point, we believe that this direction of research holds promise of revealing new ways that entertainment can be particularly meaningful for individuals, and may also represent a promising avenue for harnessing positive media influence.

CONCLUDING THOUGHTS

As we write this chapter, it is likely that newer forms of entertainment have begun to emerge and have already started to replace the ones discussed here. The use of the word "evolving" seems almost odd because it connotes gradual change, when technologies are actually developing at a dizzying pace. In this chapter, we suggest that these technological changes shape how we consume entertainment, what constitutes entertaining content, and even how we perceive and structure the world through the entertainment lens that is now constantly embedded in our social and emotional lives. Entertainment and "real life" are perhaps blurred now more than ever before.

We are concerned about how this blurring may threaten to trivialize and devalue many aspects of life that we normally hold dear—the beauty of a sunset, a baby's first steps, or the reunion of old friends. In this respect, we acknowledge and applaud what we perceive to be a growing "slow-technology movement" where people are recognizing the importance of putting away their technologies so as to fully appreciate the meaningfulness of the present. At the same time, we also recognize the potential of the "reality" inherent in newer forms of entertainment to give us the opportunity to experience a full range of social emotions, and to get glimpses of actual displays of love, compassion, and kindness in wide-ranging and diverse populations. In this respect, we are hopeful and excited about the idea that emerging technologies may hold promise of enhancing our connectedness and our awareness of shared human values.

Of course, our arguments about the dangers and the potentials of entertainment technologies to deliver meaningful experiences await our future investigation. Perhaps like all newer types of entertainment that have evolved throughout history, the ones discussed in this chapter will have both positive and negative ramifications. As media scholars, though, it is our important calling to be both critics and advocates, and to do so in ways that enable present and future audiences to richly experience the meaningful and deeply gratifying insights afforded by life with and without media entertainment.

REFERENCES

60 Minutes/Vanity Fair Poll: Etiquette. (2014, July 7). *CBS News*. Retrieved from http://www. cbsnews.com/news/60-minutes-vanity-fair-poll-etiquette/

Algoe, S. B., & Haidt, J. (2009). Witnessing excellence in action: The "other-praising" emotions of elevation, gratitude, and admiration. *The Journal of Positive Psychology, 4*(2), 105–127. doi: 10.1080/17439760802650519

Aristotle. (trans. 1931). *Nicomachean ethics* (W. D. Ross, Trans.). London: Oxford University Press.

Busselle, R., & Bilandzic, H. (2008). Fictionality and perceived realism in experiencing stories: A model of narrative comprehension and engagement. *Communication Theory, 18*(2), 255–280. doi: 10.1111/j.1468-2885.2008.00322.x

Cantor, J. (2009). Fright reactions to mass media. In J. Bryant & M. B. Oliver (Eds.), *Media effects: Advances in theory and research* (3rd ed., pp. 287–303). New York: Routledge.

Corey Ann. (2013, May 24). Why you might want to consider an unplugged wedding. *The Huffington Post*. Retrieved from http://www.huffingtonpost.com/bridal-guide/why-you-might-want-to-con_b_3331528.html

deGuzman, C. (2013). I forgot my phone [Video file]. Retrieved from https://www.youtube.com/watch?v=OINa46HeWg8

Freeman, D., Aquino, K., & McFerran, B. (2009). Overcoming beneficiary race as an impediment to charitable donations: Social dominance orientation, the experience of moral elevation, and donation behavior. *Personality and Social Psychology Bulletin, 35*(1), 72–84. doi: 10.1177/0146167208325415

Grizzard, M., Tamborini, R., Lewis, R. J., Wang, L., & Prabhu, S. (2014). Being bad in a video game can make us more morally sensitive. *Cyberpsychology, Behavior, and Social Networking, 17*(8): 499–504. doi:10.1089/cyber.2013.065

Haidt, J. (2003). The moral emotions. In R. J. Davidson, K. R. Scherer, & H. H. Goldsmith (Eds.), *Handbook of affective sciences* (pp. 852–870). Oxford: Oxford University Press.

Hartmann, T., Toz, E., & Brandon, M. (2010). Just a game? Unjustified virtual violence produces guilt in empathetic players. *Media Psychology, 13*(4), 339–363. doi: 10.1080/15213269.2010.524912

Henkel, L. A. (2014). Point-and-shoot memories: The influence of taking photos on memory for a museum tour. *Psychological Science, 25*(2), 396–402. doi: 10.1177/0956797613504438

Hollenbaugh, E. E., & Ferris, A. L. (2014). Facebook self-disclosure: Examining the role of traits, social cohesion, and motives. *Computers in Human Behavior, 30*, 50–58. doi: 10.1016/j.chb.2013.07.055

Kid President. (2013, January 4). A pep talk from Kid President to you. [Video file]. Retrieved from http://youtu.be/l-gQLqv9f4o

Klimmt, C., & Vorderer, P. (2009). Media entertainment. In C. R. Berger, M. E. Roloff, & D. Roskos-Ewoldsen (Eds.), *The handbook of communication science* (pp. 345–361). Thousand Oaks, CA: Sage.

Klimmt, C., Vorderer, P., & Ritterfeld, U. (2007). Interactivity and generalizability: New media, new challenges. *Communication Methods and Measures, 1*(3), 169–179. doi: 10.1080/19312450701434961

Lee, K. M. (2004). Presence, explicated. *Communication Theory, 14*(1), 27–50. doi: 10.1093/ct/14.1.27

Lewis, R. J., Tamborini, R., & Weber, R. (2014). Testing a dual-process model of media enjoyment and appreciation. *Journal of Communication, 64*(3), 397–416. doi: 10.1111/jcom.12101

Marshall, N. (2013, October 8). Gravity: I love you George Clooney but you make me sick. *The Guardian*. Retrieved from http://www.theguardian.com/culture/2013/oct/09/gravity-i-love-you-george-clooney-but-you-make-me-sick

Meeker, M., & Wu, L. (2013, May 29). 2013 Internet trends. *Kleiner Perkins Caufield & Byers*. Retrieved from http://www.kpcb.com/insights/2013-internet-trends

Mendelson, A. L., & Papacharissi, Z. (2011). Look at us: Collective narcissism in college student facebook photo galleries. In Z. Papacharissi (Ed.), *A networked self: Identity, community, and culture on social network sites* (pp. 251–273). New York: Routledge.

Misra, S., Cheng, L., Genevie, J., & Yuan, M. (2014). The iPhone effect: The quality of in-person social interactions in the presence of mobile devices. *Environment & Behavior*, Advance online publication. doi: 10.1177/0013916514539755

Nabi, R. L., Biely, E. N., Morgan, S. J., & Stitt, C. R. (2003). Reality-based television programming and the psychology of its appeal. *Media Psychology, 5*, 303–330. doi: 10.1207/S15327 85XMEP0504_01

Oliver, M. B. (1994). Portrayals of crime, race, and aggression in reality-based police shows: A content analysis. *Journal of Broadcasting & Electronic Media, 38*(2), 179–192. doi: 10.1080/08838159409364255

Oliver, M. B. (2009). Entertainment. In R. L. Nabi & M. B. Oliver (Eds.), *The Sage handbook of media processes and effects* (pp. 161–175). Thousand Oaks, CA: Sage.

Oliver, M. B., Ash, E., Kim, K., Woolley, J. K., Hoewe, J., Shade, D. D., & Chung, M.-Y. (in press). Media-induced elevation as a means of enhancing feelings of intergroup connectedness. *Journal of Social Issues*.

Oliver, M. B., Ash, E., & Woolley, J. K. (2013). The experience of elevation: Responses to media portrayals of moral beauty. In R. Tamborini (Ed.), *Media and the moral mind* (pp. 93–108). New York: Routledge.

Oliver, M. B., Ash, E., Woolley, J. K., Shade, D. D., & Kim, K. (2014). Entertainment we watch, and entertainment we appreciate: Patterns of motion picture consumption and acclaim over three decades. *Mass Communication & Society, 17*(6), 853-873. doi: 10.1080/15205436.2013.872277.

Oliver, M. B., & Bartsch, A. (2011). Appreciation of entertainment: The importance of meaningfulness via virtue and wisdom. *Journal of Media Psychology: Theories, Methods, and Applications, 23*(1), 29–33. doi: 10.1027/1864–1105/a000029

Oliver, M. B., Hartmann, T., & Woolley, J. K. (2012). Elevation in response to entertainment portrayals of moral virtue. *Human Communication Research, 38*(3), 360–378. doi: 10.1111/j.1468–2958.2012.01427.x

Oliver, M. B., & Raney, A., (Eds.). (2014). Broadening the boundaries of entertainment research [Special issue]. *Journal of Communication, 64*, 361–568.

Ophir, E., Nass, C., & Wagner, A. D. (2009). Cognitive control in media multitaskers. *Proceedings of the National Academy of Sciences, 106*(37), 15583–15587. doi: 10.1073/pnas.0903620106

Pew Research Internet Project. (2014). Mobile technology fact sheet. *Pew Research Center*. Retrieved from http://www.pewinternet.org/fact-sheets/mobile-technology-fact-sheet/

Phone stack. (2014). In *Urban Dictionary*. Retrieved from http://www.urbandictionary.com/define.php?term=Phone+Stack

Playing for Change. (2008, Nov. 6). Stand by me. [Video file]. Retrieved from http://youtu.be/Us-TVg40ExM

Postman, N. (1986). *Amusing ourselves to death: Public discourse in the age of show business*. New York: Penguin Books.

Przybylski, A. K., & Weinstein, N. (2013). Can you connect with me now? How the presence of mobile communication technology influences face-to-face conversation quality. *Journal of Social and Personal Relationships, 30*(3), 237–246. doi: 10.1177/0265407512453827

Raney, A. A. (2006). The psychology of disposition-based theories of media enjoyment. In J. Bryant & P. Vorderer (Eds.), *Psychology of entertainment* (pp. 137–150). Mahwah, NJ: Erlbaum.

Raney, A. A., & Oliver, M. B. (2014). Expanding the boundaries of entertainment research: An epilogue. *Journal of Communication, 64*(3), 566–568. doi: 10.1111/jcom.12091

Reinecke, L., Vorderer, P., & Knop, K. (2014). Entertainment 2.0? The role of intrinsic and extrinsic need satisfaction for the enjoyment of Facebook use. *Journal of Communication, 64*(3), 417–438. doi: 10.1111/jcom.12099

Schnall, S., Roper, J., & Fessler, D. M. T. (2010). Elevation leads to altruistic behavior. *Psychological Science, 21*(3), 315–320. doi: 10.1177/0956797609359882

Siegel, J., Dubrovsky, V., Kiesler, S., & McGuire, T. W. (1986). Group processes in computer-mediated communication. *Organizational Behavior and Human Decision Processes, 37*(2), 157–187. doi: 10.1016/0749–5978(86)90050–6

Silvers, J. A., & Haidt, J. (2008). Moral elevation can induce nursing. *Emotion, 8*, 291–295. doi: 10.1037/1528–3542.8.2.291

Sky News. (2013, May 21). Oklahoma tornado: Dog emerges from debris. [Video file] Retrieved from http://youtu.be/BUc1ZgXiQGs

Sweet, N. G. (2013, September 30). Put that away! Phone stacking and other solutions to your phone addiction. *KQED.* Retrieved from http://blogs.kqed.org/pop/2013/09/30/put-that-away-phone-stacking-and-other-solutions-to-your-phone-addiction/

Tamborini, R. (2011). Moral intuition and media entertainment. *Journal of Media Psychology: Theories, Methods, and Applications, 23*(1), 39–45. doi: 10.1027/1864–1105/a000031

Tamborini, R. (2013). A model of intuitive morality and exemplars. In R. Tamborini (Ed.), *Media and the moral mind* (pp. 43–74). New York: Routledge.

Tan, E. S. (1996). *Emotion and the structure of narrative film: Film as an emotion machine.* Mahwah, NJ: Erlbaum.

Turkle, S. (2011). *Alone together: Why we expect more from technology and less from each other.* New York: Basic Books.

Vandewater, E. A., Shim, M., & Caplovitz, A. G. (2004). Linking obesity and activity level with children's television and video game use. *Journal of Adolescence, 27*(1), 71–85. doi: 10.1016/j.adolescence.2003.10.003

Vorderer, P., & Kohring, M. (2013). Permanently online: A challenge for media and communication research. *International Journal of Communication, 7*, 188–196.

Zillmann, D. (2000). Mood management in the context of selective exposure theory. In M. E. Roloff (Ed.), *Communication yearbook* (Vol. 23, pp. 103–123). Thousand Oaks, CA: Sage.

Zillmann, D., & Vorderer, P. (Eds.). (2000). *Media entertainment: The psychology of its appeal.* Mahwah, NJ: Erlbaum.

CHAPTER FIVE

Media Policy for Happiness

A Case Study of Bhutan

PENG HWA ANG
NANYANG TECHNOLOGICAL UNIVERSITY, SINGAPORE

Probably more than any other country, Bhutan has taken a deliberate and cautious path to development. The notion of Gross National Happiness (GNH), perhaps Bhutan's best-known intellectual gift to the rest of the world, is also the best example of how the country sees the purpose of development—not merely as an economic pursuit, but more a self-actualization of individuals in the collective. The premise is that the ultimate aspiration of every human being is happiness.

This is not far from the Declaration of Independence of the United States, which holds it self-evident that among the inalienable rights of all people is "the pursuit of happiness." The major difference is that Bhutan has embedded GNH into its national constitution and established institutions and policies to operationalize it. Hence, policies for media development are viewed through the perspective of GNH principles. This is unique in the world. On the one hand, it offers exciting possibilities of "thinking out of the box" and exploring alternative paths of developing the media. On the other hand, as a unique case, such a notion is as challenging as it is trailblazing; there is no example to follow.

Notwithstanding these challenges, the Bhutanese government wants GNH applied to preserve the uniqueness of Bhutan and its values such as the average Bhutanese should not be materialistic and he/she should not blindly ape what is portrayed on TV. And the government expects media to play a watchdog role in ensuring good governance of the country (Dorji, n.d.). Bhutan, therefore, offers an instructive answer to the question: Can media policy be oriented to achieve goals of well-being and happiness?

THE BHUTAN CASE

In order to answer this question, we have to begin at the end—what is the purpose of media? In this chapter, media is taken as a collective and so the singular tense will be used. This case study was also part of a larger project looking at the media policies of emerging democracies. Conventionally in modern participatory democracies, media is expected to play the role of a watchdog for the society and thereby to enhance the democratic process. In order for media to perform that role, it has to be independent and free. This means being independent of government and business influence. In a free-market capitalist model, which theory seems to suggest to be a preferred model, it means that media must be financially independent, even as it depends on advertising from business to be financially viable.

However, media products—which are supposed to enhance democracy—sell the best and are most profitable for the media organization when they appeal to the largest audiences. Often, the audience is more interested in sports, entertainment, and scandals rather than politicians debating over serious issues. It is therefore evident that the business model is fraught with conflicts and tensions. It is into the fray that media law and policy wade to mitigate if not resolve the conflicts and tensions. To ensure that news reports are truthful and robust so as to attract both readers and advertisers, there is a bright line between editorial and advertising. To ensure that the media can be an effective watchdog, there are rules promulgated by the government to protect it from political and financial interference.

It was with a keen awareness of these tensions that the Kingdom of Bhutan approached media policy, when the king stepped down to transform the kingdom into a democracy. Unlike other countries that achieved democracy through a bottom-up revolution, Bhutan's democracy was "granted" by the 4th King Jigme Singye Wangchuck. He said, "To ensure long-term happiness for Bhutanese people we must promote democracy, because an effective political system is more important than a throne" (Fan, 2008). Where the king had provided good governance, today a democratically elected government does so. The premise of His Majesty's perception was that a small, and therefore vulnerable, society should not be left to the hands of one individual (Fan, 2008).

Until the late 1990s, landlocked Bhutan was so isolated that it had just one radio station and one newspaper, both of which were owned by the King, but no television or Internet. Then, in rapid development, Bhutan made the transition from an absolute monarchy to a democratic constitutional monarchy with the first general election in 2007 and a general liberalization of the government-dominated economy (Turner, Chuki, & Tshering, 2011).

Key to guiding development is the notion of GNH. This was not intended to replace Gross National Product (GNP) but was instead intended as a concept and philosophy to build the national identity of the Bhutanese. The country, tiny by population, is sandwiched between the two most populous countries in the world—China and India. GNH was intended to assert the Bhutanese identity even as the country opened up to the world so as to resist the onslaught of the Chinese and Indian cultures, values, and lifestyles. Identity is asserted most visibly through attire and architecture. The men's *gho* and the women's *kira* hail from the 15th century and are worn by school children and civil servants; all houses must have carved windows and an orange-color motif at the eaves. But the Bhutanese have given the matter much thought. They also want the notion of GNH and their identity asserted through all aspects of Bhutanese life, including media (Dorji, n.d.).

DEVELOPMENT OF THE GNH INDEX

In asserting its identity, Bhutan treats cultural distinctiveness as being of intrinsic value, useful for the preservation of the sovereignty of a nation faced with asymmetry of power with its neighbors (Ura, 2004). In the words of the President for the Centre for Bhutan Studies, "Culture is cultivated and revived as an anchor in a sea of change" (Ura, 2004, p. 6). Implicit in this is a shared consciousness of what it means to be Bhutanese. That is, the Bhutanese society must share a common set of values that are often instilled by the media. Reading, listening, and watching the same media can reinforce the shared values, provided that media portray more of those Bhutanese values. And GNH is the one overarching value.

Defining GNH

The term GNH was officially coined by Bhutan's 4th King Jigme Singye Wangchuck in 1979, who said that Gross National Happiness was more important than Gross National Product (Ura, Alkire, Zangmo & Wangdi, 2012a, p. 6). Initially viewed as a casual remark, the GNH rang sincere as an aspiration to combine the country's unique culture of Buddhist values with the goals of socio-economic development. Thus, the Bhutanese have defined GNH as:

> Gross National Happiness measures the quality of a country in a more holistic way [than GNP] and believes that the beneficial development of human society takes place when material and spiritual development occur side by side to complement and reinforce each other (Ura, Alkire, Zangmo, & Wangdi, 2012b, p. 7).

GNH is a comprehensive concept that has been explicated by the Centre for Bhutan Studies. Among Buddhists, "happiness" has a much broader meaning than what is taken for granted in the Western literature. Unlike GNP, which only emphasizes economic growth, GNH aims to consider development from a holistic perspective, taking into account the state as well as individual development in the collective. Instead of unbalanced development under GNP, development under GNH aims to balance between modernization and tradition and between materialism and spiritualism. It aims at a holistic approach toward progress based on both economic and non-economic factors. The focus is on human well-being (Ura, 2004).

In Bhutan, GNH is not merely a development philosophy but a constitutional pledge. The new 2008 Constitution in Article 9 states: The State shall strive to promote those conditions that will enable the pursuit of Gross National Happiness. This means that unlike most other countries, which may have increased GNP but at the expense of their cultures, environment, and their traditional values, Bhutan aims at promoting economic development without diluting its culture, social systems, and spiritual values (Ura & Galay, 2004). The current and 5th King of Bhutan, HM Jigme Khesar Namgyel Wangchuck, said that fulfilling the vision of GNH would be one of the four main responsibilities of his reign (UNDP, 2011).

GNH was popular nationwide, and also attracted the attention of scholars from different disciplines and different parts of the world. Several international conferences on GNH have been held, beginning with the first in 2004 to operationalize the concept and apply it in governance (see Table 5.1).

Table 5.1. International Conferences on Gross National Happiness.

	Date	Venue	Theme
1.	Feb 2004	Thimphu, Bhutan	1st International Conference on GNH: Operationalizing GNH
2.	June 2005	Nova Scotia, Canada	2nd International Conference on GNH: Rethinking Development: Local Pathways to Global Wellbeing
3.	Nov 2007	Bangkok, Thailand	3rd International Conference on GNH: Towards Global Transformation
4.	Nov 2008	Thimphu, Bhutan	4th International Conference on GNH: GNH: Practice and Measurement
5.	Aug 2009	Iguacu, Brazil	5th International Conference on GNH: GNH in Practice
6.	Apr 2012	New York City, USA	UN Conference Wellbeing and Happiness: Defining a New Economic Paradigm

Measuring Happiness

It should be noted that there are two major approaches to measuring "happiness" at the individual level and neither was adopted for use in Bhutan. One approach uses the Positive and Negative Affect Scale (Watson & Clark, 1994) and links positive and negative affect to other assessments such as the Satisfaction with Life scale (Diener, Emmons, Larsen, & Griffin, 1985). The other is based on the more recent field of positive psychology (Seligman, 2002), which assesses individual strength, character virtues, and flow. The Centre for Bhutan Studies did not adopt either approach, choosing instead to develop its own.

With some external intellectual assistance, the Centre for Bhutan Studies developed a sophisticated instrument of social survey to measure well-being, the first of which was released in 2008. To develop the GNH indicators, the Centre for Bhutan Studies conducted a pilot survey of 350 respondents from remote, semi-urban, and urban populations. The survey, funded in part by the UN Development Programme, took three months to complete with the first interviews taking seven to eight hours. In the words of the President of the Centre for Bhutan Studies, "The pilot survey questionnaire, which was found to be too lengthy, was pared down to a questionnaire that took half a day to interview in the final survey...." (Ura, n.d.). In the final survey, a total of 950 respondents were interviewed for half a day for 758 variables each over four months from December 2007 to March 2008.

The sophistication of the GNH Index is demonstrated in how the indicators were selected:

> The GNH measure has been designed to include nine core *domains* that are regarded as components of happiness in Bhutan and is constructed of *indicators*, which are robust and informative with respect to each of the domains. The nine domains were selected on *normative* as well as *statistical* grounds, and are equally *weighted*, because each domain is considered to be relatively equal in terms of its intrinsic importance as a component of gross national happiness. Within each domain, two to four *indicators* were selected that seemed likely to remain informative across time, had high response rates, and were relatively uncorrelated (Centre for Bhutan Studies, n.d.; emphasis was in the original text).

It can be seen, therefore, that the Bhutanese have faced what researchers face—the tension between being as accurate and comprehensive as possible and the real-world difficulties in getting the data to fulfill that desire.

Table 5.2 shows the nine core domains and their 33 indicators. Some of the dimensions cover the traditional concerns of development such as living standard, health, and education. Some newer dimensions such as time use, good governance, ecological diversity and resilience were added. In 2010, the number of indicators was reduced from 72 to 33 under the same nine domains based upon a robust multi-dimensional methodology known as the Alkire-Foster (Alkire, 2007) method. These included psychological well-being, health, education, time use, cultural diversity, resilience, and living standards. The full survey questionnaire is available at http://www.grossnationalhappiness.com. These indicators have been used to help ensure that government policies are aligned with GNH (Centre for Bhutan Studies, 2012).

Table 5.2. GNH Domains, Indicators, and Types of Well-Being.

Domains	Indicators	Type of Well-Being
1. Psychological well-being	General psychological distress	EvWB
	Emotional balance	ExWB
	Spirituality	EuWB
2. Time use	Sleeping hours and total working hours	EvWB
3. Community vitality	Family vitality	ExWB
	Safety	EvWB
	Reciprocity	EuWB
	Trust	
	Social support	
	Socialization	
	Kinship density	
4. Health	Health status	ExWB
	Health knowledge	EvWB
	Barrier to health	
5. Education	Education attainment	EuWB
	Dzongkha language	
	Folk and historical literacy	
6. Cultural diversity and resilience	Dialect use	EvWB
	Traditional sports	EuWB
	Community festival	
	Artisan skill	
	Value transmission	
	Basic precept	
7. Ecological diversity and resilience	Ecological degradation	
	Ecological knowledge	
	Afforestation	
8. Living standard	Income	ExWB
	Housing	EvWB
	Food security	
	Hardship	
9. Good Governance	Government performance	
	Freedom	EuWB
	Institutional trust	

Note. Source of NGH domains and indicators is Ura (u.d.). The last four domain areas are the areas being focused on by the government.

The GNH Index for Bhutan was calculated as 1 minus the product of two measures HA (GNH = 1-HA), where H (for *headcount*) is the percentage of those who do not enjoy sufficiency in six or more domains and A is the *average proportion* of domains in which people who are not happy lack sufficiency and is intended to capture the breadth of shortfalls (Centre for Bhutan Studies, n.d.).

The 2010 survey of 6,467 respondents yielded the following results: (1) Headcount was 40.8% who may be considered "happy" because they have achieved sufficiency in six or more of the nine domains, with the remainder 59% of Bhutanese either "narrowly happy or unhappy." (2) Intensity was 43.4%, which means the 59% of Bhutanese who were not considered "happy" lack sufficiency in 43% of the domains. Nine domains times 0.43 is 3.87. Thus unhappy Bhutanese on average lack sufficiency in just under four domains and enjoy sufficiency in just over five domains. (3) Therefore, GNH Index = 0.743 = 1–0.591x0.434. The GNH Index ranges from 0 to 1, the higher the better. It reflects the percentage of Bhutanese who are happy and the percentage of domains in which not-yet-happy people have achieved sufficiency (Gross National Happiness Commission, 2009).

Notwithstanding the survey, the 10th Five-Year Plan of Bhutan (2008–2013) has simplified the nine domain areas and focused on four as pillars of GNH: sustainable and equitable socioeconomic development, environmental conservation, the preservation and promotion of culture, and good governance (Gross National Happiness Commission, 2009).

GNH and Types of Well-Being

It should be evident that the GNH is different from psychological measures of subjective well-being (SWB). Research in that field has developed multi-faceted dimensions of SWB that now include evaluative well-being, experienced well-being, and eudaimonic well-being. *Evaluative well-being* (EvWB) refers to judgments of how satisfying one's life is; these judgments are sometimes applied to specific aspects of life, such as relationships, community, health, and work. *Experienced well-being* (ExWB) is concerned with people's emotional states and may also include effects associated with sensations (e.g., pain, arousal) and other factors such as feelings of purpose or pointlessness that may be closely associated with emotional states and assessments of those states. ExWB is often further divided into positive experiences, which may be characterized by terms such as joy, contentment, and happiness, and negative experiences, which may be characterized by sadness, stress, worry, pain, or suffering. *Eudaimonic well-being* (EuWB) refers to a person's perceptions of meaningfulness, sense of purpose, and the value of his or her life (Stone & Mackie, 2013).

Table 5.2 categorizes the indicators into one of the types of well-being. The most frequently appearing is EvWB, a judgment of satisfaction in life. This makes sense as the index is not so much about the emotional state of ExWB and is even less about the happiness of finding purpose in life in EuWB. It may, however, also be an artifact of a survey instrument: The questions are of an evaluative nature and less about capturing subjective happiness of ExWB. Table 5.2 shows that not all the indicators can be placed into one of the three categories, and in fact the entire domain area of "ecological diversity and resilience"—which measures conservation—is not even in any of the three categories. This emphasizes the point that the GNH is not about happiness in the psychological sense although there are clear overlaps with SWB. With this broad perspective of the GNH index, this case study focused on the role of media policies as to how they might help achieve the goals of GNH.

METHOD

I spent a total of three weeks in Bhutan over two visits in 2012 as part of a consultancy project with the Bhutanese Government to review the outcome of the country's media policies. The laws had been implemented in 2008 by the first democratically-elected parliament. I was given access to materials related to the personnel involved with the media. The key informant interviews ranged from one to two hours and the interviewees included: The head of the media regulatory agency, the Bhutan Information, Communication and Media Authority (BICMA); the head of the first private radio station Kuzoo, which was written about in the book *Radio Shangri-La* (Napoli, 2012); editors of leading newspapers the *Kuensel, Bhutan Observer*, the *Bhutanese, Bhutan Times, Bhutan Today*; the heads of Bhutan Broadcasting System; the head of the Motion Picture Association of Bhutan; and officials at the Ministry of Information and Communications.

The Context

Bhutan has a population of about 750,000 spread over an area so vast that sometimes one family lives on one mountain with the nearest neighbor on the next mountain. This means the media market is small and spread out. The result is that some media businesses, such as newspapers, find distribution of their products particularly challenging. The largest city is the capital Thimphu, which has a population of around 125,000 or about a sixth of the total population of the country. Partly because of the larger population and the greater literacy in the capital, many of the country's media organizations and outlets are concentrated there. In 2012, when I visited Thimphu, the country had 12 newspapers with one daily, a couple of bi-weekly, and the rest appearing weekly. It was probably the city with the most newspapers on the planet.

Role of Media in GNH

All media operate in a context. For Bhutan, the GNH-inspired vision and policies and values that are meant to guide all areas of development, and the nature of Bhutanese society are of great importance in determining the direction of development of the media sector. Because GNH is the bedrock of all policies in Bhutan, this means that the media policy must support the values of GNH and its vision of democracy and human well-being and not pursue commercialism for its own sake.

As the four pillars of the GNH—sustainable and equitable socio-economic development, environmental conservation, the preservation and promotion of culture, and good governance—are intended to be pursued as goals in all walks of Bhutanese life, this means that the policies and structures in the media industry, too, must be sustainable in the long run, help preserve and promote Bhutanese culture, help promote the conservation of the environment, and ensure good governance.

Media is mandated to help develop society, not just themselves. The media, however, is not itself expected to be "GNH media," which for a while was thought to mean that published news should not make readers unhappy. This would also have meant not criticizing the government because doing so would make some officials unhappy. Over time, it has come to be understood that attaining GNH meant aiming to fulfill the four pillars of GNH. The next section aims to address the question of how the media may do so.

ATTAINING GNH BY BHUTANESE MEDIA

To understand media's role with respect to GNH, one has to begin first with understanding the peculiarities of the media in Bhutan. Bhutanese media has grown by leaps and bounds in the last few years. *Kuensel* was the only national newspaper of Bhutan until 2006 when *Bhutan Times* and *Bhutan Observer* were launched. Bhutan Broadcasting Service launched its first radio broadcast in 1979 and its first television channel in 1999. With liberalization, however, by September 2012, Bhutanese media included 12 newspapers, four magazines ranging from entertainment to news, six radio stations, about 40 television channels (mainly international) and three Internet Service Providers both privately- and government-owned. Table 5.3 tabulates the mass media landscape in Bhutan. It shows that the media, especially print media, is quite diverse in Bhutan. Although most of the media organizations are privately owned, a large share of the value is held by government entities.

Table 5.3. Media in Bhutan.

Media	Private Owned	Year Started	Government / Public Owned	Year Started	Total
Newspaper	Bhutan Times	2006	Kuensel	1967	12
	Bhutan Observer	2006			
	Bhutan Today	2008			
	Business Bhutan	2009			
	The Journalist	2009			
	Druk Nyetshul	2011			
	Druk Yoedzar	2011			
	Bhutan Youth	2011			
	The Bhutanese	2012			
	Gyalchi Sharchog	2012			
	Druk Melong	2012			
Radio	Kuzoo FM	2006	Bhutan Broadcasting Service (BBS)	1973	6
	Radio Valley	2007			
	Centennial Radio	2008			
	Radio High	2010			
	Radio Waves	2010			
Television			BBS TV	1999	1
Internet Service Providers	Drukcom	2004			3
	Samden	2004			
	Tashi Cell	2006			
Magazines	Druk Trowa	2009			4
	Drukpa	2009			
	Bhutan Window	2009			
	Yeewong	2009			

The dominance of government-owned media is illustrated in Table 5.4. Only four of the 11 privately-owned newspapers submitted themselves for a media audit and could be audited. Three others could not be audited because of the unavailability of records. Table 5.4 shows that the government-owned English-language *Kuensel* is far and away the leading newspaper, with a circulation that beat the next four combined (7,159 vs 5,151).

Table 5.4. Audited Circulation of Newspapers in Bhutan 2013.

Newspaper	Frequency	Ownership	Circulation Per Issue Hardcopy	PDF
Kuensel	Daily	Government	7,159	1,541
Kuensel (Dzonghka ed.)	Daily	Government	1,306	X
The Bhutanese	Weekly	Private	1,712	69
The Journalist	Weekly	Private	1,300	X
Bhutan Today	Weekly	Private	1,291	58
Business Bhutan	Weekly	Private	848	X

Note. Source: Dorji (2014).

The Bhutanese Government, cognizant of the infancy of its democracy, had instituted policies that favored new entrants in the media industry. Capital grants, particularly in training, have been awarded in favor of entrants. While the open licensing policy guaranteeing an open platform for different players to compete with each other in a free market has brought about media diversity and plurality, the circulation figures suggest that the competition is not sustainable (Ministry of Information and Communications, 2010). In fact, industry players for the most part seem cognizant of this and had resisted a media audit. At the time of my visit in 2012, all the newspapers except the dominant *Kuensel* were unprofitable. The situation was clearly not sustainable.

Sustainable and Equitable Socio-Economic Development

The biggest issue facing the media in Bhutan is sustainability. In the initial years of media liberalization, content was focused more on information and education, which was highly appreciated then (Department of Information and Media, 2008). With new entrants into the media, the function of education has been blurred. Media, under the pressure of competition to capture more viewers, listeners, and readers, has gradually devoted more space to entertainment.

The officials hope that media content in Bhutan would be more thoughtful with less of crass commercialism and media should also not encourage consumerism or the blind aping of behavior and culture. The reality though is that most Bhutan media companies were unprofitable. The main reason is that the advertising pie in Bhutan is small with the government accounting for 70% to 80% of it.

The heavy reliance on government as the source of advertising revenue raises questions about the extent to which the media can further democracy through their role as a watchdog of government. This heavy reliance on government advertising is a key reason Bhutan media is ranked "partly free" by Freedom House. At least one news report suggests that a politician or bureaucrat unhappy with a story can influence the placement of advertisements (*The Bhutanese*, 2012).

The reliance on government advertising also means that the growth in advertising will have to depend on the growth of government revenue. As government revenues come from taxes, for advertising to increase, taxes will have to increase. This is unlike the case of advertisements in the private sector where more advertising can translate into more revenue, at least theoretically, which can then lead to more advertising. Also, the private sector sometimes advertises more in response to advertising by its competitors.

In more developed countries, some of the tender notices that currently appear in newspapers and television in Bhutan, are disseminated in the form of email notices and online alerts as these can be more efficient and cost-effective when the target audience of vendors is known. In short, the Bhutan advertising pie in its current form does not bode well for the media industry and is therefore not likely to be sustainable.

The media's reliance on the government is becoming a mutually reinforcing interdependency. The media is becoming politicized as politicians learn to play to the media, such as not to antagonize the media in the run-up to the elections. A code on advertising placement that would have placed government advertising based on circulation and reach was ready in 2012 but was not implemented until after the 2013 elections. Such interdependency alters the relationship between politicians and media owners and journalists. It blunts the media's ability to be a watchdog.

The media industry, particularly that for newspapers, suffers from a lack of professionally trained staff. It is a vicious cycle: As the newspapers are intensely competitive and unprofitable, they are not able to pay well, which means they are not able to attract good journalists, which affects the quality of the output, which in turn affects circulation. The long-term solution is for the small and unprofitable newspapers to close so as to allow some better ones to survive. A shake-out began soon after the 2013 elections (*The Bhutanese*, 2014). In all likelihood, more newspapers will close over time.

Preservation and Promotion of Culture

As discussed earlier, Bhutanese cultural distinctiveness is critical to its identity. The distinctiveness is most evidently displayed in the spiritual heritage, dress, language, and creative industry. It also extends, less immediately visible, to the arts. The main strategy to preserve and promote local culture has been to produce local content. Media attempts to play a role in helping with the appreciation of Bhutanese culture. Traditional music and dance are broadcast on TV in order to reinforce the national culture. Given the small media market, it is cheaper to import foreign content, especially television and movies, than to produce them domestically. Indian and Western popular cultures are becoming more prevalent, particularly among the young (Department of Information and Media, 2008). Indian television programming with Bollywood content is drawing an increasing number of viewers, making the viability of domestic TV competition questionable. Such foreign content tends to have high production quality, making them more visually appealing and therefore a greater threat of "diluting" the culture. Although a 2008 study by the Centre for Media and Democracy found that Bhutanese preferred their own television station (29.5%) to Indian channels (22.3%), the concern nevertheless is that the Indian programs will have sway over the Bhutanese, particularly the young (Department of Information and Media, 2008).

In line with the policy of preserving its culture, all media has to carry some content in Dzongkha. Newspapers have to carry at least one page of the Dzongkha language in order to obtain government advertising. Some complained that their competitors were taking shortcuts in simply lifting pages off the Internet. A logical step would be to impose quotas on the imports of such media, particularly when they are of foreign origin, while incentivizing media organizations to create more local content.

Conservation of the Environment

As a country that generates power through renewable hydro-electricity, Bhutan is one of the few countries in the world with net greenhouse gas sequestration capacity (UNDP, 2010). Climate change, however, is already affecting the glaciers, which produces the run-off to power the hydro-electric dams that in turn generate income for the country. Officials suggested that media could play an important role in raising awareness and educating people on Bhutan's location in the fragile eco system. It has been suggested, and various stakeholders are exploring, the possibility of sharing common communication infrastructure by business rivals both to conserve the environment as well as minimize wasteful duplication of resources. Such infrastructure may include printing plants, mobile phone towers, and broadcast towers.

Governance: Watchdog Role

Of the four pillars, the most important and most debated is governance and the media's role in it. There are two aspects to the issue. The first is the traditional view of the media as a watchdog of wrongdoings in society; the second is the governance of the media itself. The watchdog role is not only a self-definition by the journalists, but it also reflects the expectation of the public. This role of the media is crucial to Bhutan as it attempts to achieve good governance as a pillar of GNH. To a young democracy that holds good governance as one of its four pillars, a vigilant media can help ensure the transparency and accountability of the government and thereby foster good governance. An informed, empowered, and active citizenry will be better able to nurture democracy (Voltmer, 2008).

Bhutan has a small population and "everybody knows everybody." In such a setting, notwithstanding constitutionally guaranteed rights of the media such as freedom of expression and right to information, criticisms tend to be muted and circumspect. It is important, however, for Bhutan with its goal of GNH to have a degree of robust debate while ensuring that those debating are still able to part as friends. It is important, therefore, to be analytical without being confrontational.

In playing this role, the media industry itself needs to adopt best practices in good governance. First, it has been observed by critics that media in new democracies often seem to lack the qualities that enable them to play a key role in promoting accountability and inclusive politics (Voltmer, 2008). In Bhutan's case, it may be financial ability of the media.

The media has the unquestioned role of being a watchdog of government. As a watchdog, however, media organizations themselves are expected to model good governance conduct that they expect from government, such as transparency and accountability. In this context, the heavy reliance on government advertising is not desirable in the long run. Government officials in turn should also appreciate that advertising should not be used as a "weapon" against the media, withheld when the press publishes unfavorable reports. It is a symbiotic relationship with both the government as advertiser and the media accommodating each other to perform their respective roles.

In order for media, particularly media that report the news, to perform the watchdog role, media organizations should be as independent as possible of financial, political, and other interferences of vested interests that may act against the interests of society at large. In the case of the press, journalism's first loyalty is to the larger public interest (Carey, 2002).

It would be unrealistic, however, to expect media to be completely free of such vested interests, as media will have to rely on advertising, be it from the government or from business. A viable framework for advertising should therefore be established for the financial sustainability of the media while preserving its independence.

RECOMMENDATIONS FOR MEDIA POLICY IN BHUTAN

Strengthening Media Regulator's Autonomy

Bhutan has a modern regulatory framework in which an independent regulator is established for convergent oversight of both ICT and media. The Bhutan InfoComm and Media Authority (BICMA) is Bhutan's sole regulatory body on media and communications. From its inception, it was intended that BICMA should be an autonomous and independent regulator. The independence of BICMA as ICT and media regulator has been enshrined in the Bhutan Information, Communications and Media Act 2006. The meaning of "independence," however, may have been misunderstood by some. Independence, as enshrined in Section 35 of the Act, is intended to discourage interference by "any Government official or public or private person," not the "complete insulation" of BICMA (Phuntsho, 2012). In 2012, the BICMA Director said that the Authority needed to have better transparency and accountability, empowerment, autonomy, and competency (Phuntsho, 2007).

To bolster its autonomy, independence, and competency, BICMA would need to modify the manner in which its staff is recruited. Under the current arrangement, BICMA staff are civil servants appointed under the Royal Civil Service Commission. It would be better to appoint staff, including the director, so that they are given security of tenure and yet independence from the Royal Civil Service's strictures. Greater autonomy in terms of employment conditions and benefits may be necessary in order for BICMA to hire quality staff who have the expertise to regulate the fast-moving ICT and media industry.

Improving Media Literacy

Although free education is provided in Bhutan from the primary to the tertiary level, the overall literacy of the whole population is low at about 63% (National Statistics Bureau, 2012). Media literacy, the ability to access, analyze, evaluate, and create media messages of all kinds (Media Literacy Project, 2012), is even lower. When television was introduced in 1999, Bhutanese children in particular were reported to be imitating wrestling moves in school; some adults were confused enough to write pained letters asking why grown men were beating up each other (BBC, 2004). The king has established the Bhutan Media Literacy Foundation to promote media literacy as the Internet was rolled out across the country.

Although the number of media organizations formed in Bhutan is on the steady rise, most of them target the urban population. This is understandable because of the larger population and also because of the availability of infrastructure for the media, including financing and staff. Policy initiatives as well as infrastructural investments are being developed to ensure that as much of Bhutan is covered as possible.

DISCUSSION AND CONCLUSION

Is it possible to regulate media for happiness? This case study of Bhutan shows that there are many "lesser" questions to answer before one can respond to that larger question.

The first is the definition of happiness. Of the three types of well-being (i.e., EvWB, ExWB, and EuWB), GNH as operationalized is closer to EvWB and EuWB than ExWB. Both EvWb and EuWB are more closely connected to life satisfaction and meaning and purpose of life than to emotional experience.

And while there had been an enthusiastic first attempt to capture GNH accurately, the subsequent study has simplified the data collection. In short, it has been challenging to capture GNH at the national level. Even when operationalized, how does one align well-being and national policy? In the case of Bhutan, it is the official religion of Buddhism that probably makes it easier than would be case for most other countries. Among the world's major religions, it is Buddhism that stresses contentment as a high virtue (Müller, 1881). So instead of striving to obtain wants, one should moderate wants. Buddhism's moderation of wants is a national value to the extent that Bhutanese embrace its teachings.

But as with adopting religious values, putting them into practice is a significant challenge. For policymakers, there is often no clear guidance to decide on the trade-offs between happiness and the issue to be decided even when one assumes that happiness is a socially desirable goal. For example, not all 12 newspapers can (nor should) survive in Bhutan. For long-term well-being, some of the newspapers must close. In the short-term, however, closing down the newspapers will leave a lot of unhappiness in its wake.

More fundamentally, some aspects of the media business militate against contentment: Advertisements, by their very nature, sow some seeds of dissatisfaction so that the audience will go out to buy the new product or service. Given the precarious financial state of most private media organizations, adopting the path of contentment in advertising is not a viable proposition.

There is also the criticism of paternalism. This is probably a less severe criticism in Bhutan because it is emerging from feudalism. Some have suggested that positive paternalism, i.e., policies that help individuals save money, live healthier, and make better decisions, might be more acceptable if they are presented more as encouragement toward a beneficial end (Huang, 2010).

Media policy has two foci—economic and cultural. The regulations to achieve economic ends, such as fair competition, would probably be common across countries. The governance model, for example, could probably be transplanted across countries. It is in the cultural policies that Bhutan's GNH grand dream has the most significant impact.

The goal of GNH, however, is a bold vision that has never been attempted by any country. The pursuit of GNH should be seen as an experiment in offering an alternative model of development. The Bhutanese themselves recognize it as such and are refining the measurement and policies. Given the importance of media in nation and identity building, it is only logical that the laws and policy surrounding Bhutan's media should be aligned with the national goal of GNH. From the perspective of policy, however, the Bhutan case is instructive in giving pause to considering policy for social good, as opposed to merely economic gain. To the extent that one thinks of Bhutan when one thinks of Gross National Happiness, the choice of GNH succeeds as an identity marker. The practical difficulties in developing and implementing GNH-friendly media policies mean that much work is yet to be done.

REFERENCES

Alkire, S. (2007). The missing dimensions of poverty data: Introduction to the special issue. *Oxford Development Studies, 35*(4), 347–359.

BBC. (2004, June 17). Has TV changed Bhutan? *BBC*. Retrieved from http://news.bbc.co.uk/2/hi/entertainment/3812275.stm

Carey, J. (2002). American journalism on, before, and after September 11. In B. Zelizer & S. Allan (Eds.), *Journalism after September 11*. (pp. 85–103). London: Routledge.

Centre for Bhutan Studies. (n.d.). *Gross National Happiness Index explained in detail*. Retrieved from http://www.grossnationalhappiness.com/docs/GNH/PDFs/Sabina_Alkire_method.pdf

Centre for Bhutan Studies. (2012). *An Extensive Analysis of GNH Index*. Retrieved from http://www.grossnationalhappiness.com/wpcontent/uploads/2012/10/An%20Extensive%20Analysis%20of%20GNH%20Index.pdf

Department of Information and Media. (2008). *Bhutan media impact study*. Thimphu, Bhutan: Royal Government of Bhutan.

Diener E., Emmons, R. A., Larsen, R. J., & Griffin, S. (1985). The satisfaction with life scale. *Journal of Personality Assessment, 49*, 71–75.

Dorji, G. (2014, February 8). Latest print figure circulation released. *Kuensel*. Retrieved from http://www.kuenselonline.com/latest-print-media-circulation-figures-released/

Dorji, K. (n.d.) *Media and democracy*. Permanent Secretary of Ministry of Information and Communications of Bhutan, Thimphu.

Fan, J. M. (2008, June 12). Bhutan: Democracy wins over monarchy. *China.org*. Retrieved from http://www.china.org.cn/international/news/2008–06/12/content_15772204_5.htm

Gross National Happiness Commission, Royal Government of Bhutan. (2009). *Tenth five-year plan (2008–2013)*. Retrieved from http://11rtm.gnhc.gov.bt/RTMdoc/TenthPlan_Vol1_Web.pdf

Huang, P. H. (2010). Happiness studies and legal policy. *Annual Review of Law & Social Science, 6*, 405–432.

Media Literacy Project. (2012). *Introduction to media literacy*. Retrieved from http://www.opi.mt.gov/pdf/TobaccoEd/IntroMediaLiteracy.pdf

Ministry of Information and Communications. (2010). *Media development assessment 2010*. Thimphu, Bhutan: Royal Government of Bhutan.

Müller, F. M. (Ed.) (1881). *Dhammapada: Volume X of the sacred books of the East*. (Translated from Pāli by various Oriental scholars.) Oxford: Clarendon Press. Retrieved from http://www.alanpeto.com/pdf/books/dhammapada.pdf

Napoli, L. (2012). *Radio shangri-la: What I learned in Bhutan, the happiest kingdom on Earth*. New York: Broadway Books.

National Statistics Bureau. (2012). *Key Indicators*. Retrieved from http://www.nsb.gov.bt/main/main.php

Phuntsho, S. (2012). *Improving efficiency and effectiveness of Bhutan InfoComm and Media Authority.* Unpublished paper.

Seligman, M. (2002). *Authentic happiness: Using the new positive psychology to realize your potential for lasting fulfillment.* New York: Free Press.

Stone, A., & Mackie, C. (Eds.) (2013). *Subjective well-being: Measuring happiness, suffering, and other dimensions of experience.* Washington, DC: The National Academies Press.

The Bhutanese. (2012, April 28). Controlling the media through advertisements. *The Bhutanese.* Retrieved from http://www.thebhutanese.bt/controlling-the-media-through-advertisements/

The Bhutanese. (2014, February 10). *The Bhutanese* is the highest circulated private newspaper in circulation audit. *The Bhutanese.* Retrieved from http://www.thebhutanese.bt/the-bhutanese-is-the-highest-circulated-private-newspaper-in-circulation-audit/

Turner, M., Chuki, S., & Tshering, J. (2011). Democratization by decree: The case of Bhutan. *Democratization, 18*(1), 184–210. DOI: 10.1080/13510347.2011.532626

United Nations Development Programme (UNDP). (2010). *Energy, environment and disaster management.* Retrieved from http://www.undp.org.bt/environment.htm

United Nations Development Programme (UNDP). (2011). *Bhutan national development report 2011.* Retrieved from http://www.gnhc.gov.bt/wp-content/uploads/2011/09/comined-NHDR.pdf

Ura, K. (n.d.). *Explanation of GNH Index.* Thimphu, Bhutan: Centre for Bhutan Studies.

Ura, K. (2004). The Bhutanese development story. *Centre for Bhutan Studies & GNH Research.* Retrieved from http://mms.thlib.org/typescripts/0000/0316/1651.pdf

Ura, K., & Galay, K. (2004). Preface. In K. Ura & K. Galay (Eds.), *Gross National Happiness and development: Proceedings of the first international seminar on operationalization of Gross National Happiness* (vii-xii). Thimphu, Bhutan: The Centre for Bhutan Studies.

Ura, K., & Kinga, S. (2004, May). Bhutan—Sustainable development through good governance. Proceedings from *Scaling Up Poverty Reduction: A Global Learning Process and Conference.* Shanghai. Retrieved from http://www-wds.worldbank.org/servlet/WDSContentServer/WDSP/IB/2004/12/07/000090341_20041207111905/Rendered/PDF/308210BHU0Governance01see0also0307591.pdf

Ura, K., Alkire, S., Zangmo, T., & Wangdi, K. (2012a). *A Short guide to Gross National Happiness index.* Thimphu, Bhutan: The Centre for Bhutan Studies. Retrieved from http://www.grossnational-happiness.com/wp-content/uploads/2012/04/Short-GNH-Index-edited.pdf

Ura, K., Alkire, S., Zangmo, T., & Wangdi, K. (2012b). *An extensive analysis of GNH index.* Thimphu, Bhutan: The Centre for Bhutan Studies. Retrieved from http://www.grossnationalhappiness.com/wp-content/uploads/2012/10/An%20Extensive%20Analysis%20of%20GNH%20Index.pdf

Voltmer, K. (2009). The media, government accountability, and citizen engagement. In P. Norris (Ed.), *Public sentinel: News media & governance reform* (137–162). Washington, DC: The World Bank.

Watson, D., & Clark, L. A. (1994). *The PANAS-X: Manual for the positive and negative affect schedule—expanded form.* Ames: The University of Iowa.

PART II

Perceptions, Connections, and Protection

CHAPTER SIX

Communication and Perceptions of the Quality of Life

LEO W. JEFFRES AND KIMBERLY A. NEUENDORF
CLEVELAND STATE UNIVERSITY, USA

DAVID ATKIN
UNIVERSITY OF CONNECTICUT, USA

While the question of what constitutes "the good life" has long been a topic of controversy and discussion, its roots in empirical science have a long history that can be built upon in communication. In fact, while pundits and prognosticators can debate what people should do to live a good life, their prescriptions represent their values rather than a template for society at large. And that's a worthy debate, but not the only place for contributions by scholars in communication.

This work has its origins in the 1960s, when governmental programs (e.g., the Great Society) attempted to improve the quality of people's lives, particularly in urban areas to provide them something closer to "the good life" (Dye, 2010). Toward that end, social scientists began measuring people's perceptions of the "quality of life" (QOL) and the objective and subjective influences on those perceptions (Campbell, Converse, & Rodgers, 1976). Largely limited to sociologists and then urban affairs scholars, this research stream has infrequently been visited by communication scholars to see how this discipline contributes to what can be learned. A casual review of a search for QOL studies today shows that the emphasis on urban programs has been exceeded by studies focusing on health and QOL issues, generally at the individual level (see *Journal of Happiness*). Given Maslow's (1968) hierarchy, it makes sense that starting with safety and health builds a stronger basis for a higher quality of life, after which we move up the ladder to look at relationships and actualization.

Researchers studying the "good life" recognize that people's subjective assessments of their QOL may be affected by assessments of the larger environment and its impact on them. From a bottom-up view, one's perceptions of the QOL in specific domains—family, school, culture and leisure—affects the overall assessment and level of satisfaction (e.g., Andrews & Withey, 1976; Headley, Veenhoven, & Wearing, 1991). We know there's a comparative element to this process, and comparisons are central to a discrepancy model described by Campbell et al. (1976), Inglehart and Rabier (1986), Michalos (1986) and others. They propose an aspiration-adjustment model where one's subjective well-being, or perceived QOL, reflects the gap between one's aspirations and goals and one's perceived situation. Both media and interpersonal communication provide much of the information, news and images that provoke such comparisons.

The 2014 ICA conference theme suggests that the changing communication environment has brought this into focus for scholars in the discipline. Surely each era provides a generation with certainties and uncertainties, whether that's changing social mores, economic difficulties, intriguing but potentially threatening opportunities, or disruptions to basic institutions and territorial units. And with these changes, what the "good life" means to subsequent generations evolves as well.

For example, the "good life" to American baby boomers meant moving to the suburbs, while younger adults and empty nesters today are increasingly intrigued with life in the central neighborhoods of metropolitan areas (Eisenberg, 2013; UPI, 2013). One sense of the "good life" is that sold in advertising—being healthy, wealthy, and, if not wise, fashionable, modern or hip, and eternally young (Belk & Pollay, 1985; Dittmar, 2007). Another, less celebrated vision is that captured in the notion of living the "good life" according to the "good book," whether that's Biblical or according to some other ethical tradition (Smilansky, 2012). An argument could be made that many of the conflicts in the world today center on conflicting views of the "good life" (e.g., fundamentalist religious views that are prescriptive versus more secular views that allow for individual freedoms and pursuit of aspirations that are in conflict with religious authorities).

APPROACHES TO STUDYING QOL

The "good life" is not lived in a vacuum. People live in family units, neighborhoods, and communities, supported by the larger culture and nation. In *The Great Convergence*, Mahbubani (2013) notes that the "vast majority of the world's people now have a common set of material aspirations" (p. 82), which have overcome ideological and religious differences. Thus, to the extent that much of "the good life" is "the consumer life," the consensus has reached around the world. He adds that there is a contradiction between this striving for a universal good material life and the stress it puts on the environment and sustainability, but that shouldn't detract from seeing how communication fits into the study of QOL.

There are several avenues suggested. First, since relationships are important to one's emotional health and satisfaction, interpersonal communication is important for our relationships with significant others, family, friends, acquaintances and coworkers (Lane, 2000); thus, interpersonal communication competence, family communication patterns and dynamics, and similar areas of study become relevant for studying QOL. Though contact with distant family and friends should increase our satisfaction with relationships, processes through which people are more interconnected have been operative for generations and such contact—and communication—probably is not always positive.

A generation ago, Cherry (1971) asked whether "world communication" is a threat or a promise, noting that in the last century communication linking people around the world speeded up with advancing technology and massive movements of people through two world wars and tourism. Movement on each of these dimensions has continued with growing immigration and dislocations of peoples, mass communication across almost all national and cultural boundaries, and instantaneous mediated and mobile communication through email, SMS, Skype, and social media.

A second line of inquiry is suggested by the conference call for papers. How does the changing communication environment—and its variegated options—affect people's abilities to live "the good life," to find a pace for activities, and to acquire sufficient personal space and privacy for establishing relationships and achieving economic goals? This is where inquiry about the impact of omnipresent and ubiquitous mobile communication technologies on our lives fits. Researchers have found evidence of addiction to use of mobile technology for messaging (Perry & Lee, 2007; Spiegel, 2012) as well as other communication technologies, such as the Internet and video games (Carbonell, Guardiola, Beranuy, & Bellés, 2009).

Third, communication in all of its forms serves dual functions as potential influences on people's QOL. One's personal communication network—interpersonal, mediated, organizational, mass media—can affect perceived QOL. Merely see how people react when some aspect of that network changes, when a loved one moves and is unreachable or when a favored media outlet closes. In addition, communication is the process through which we learn about the environment, impacting us directly by telling us what the options are for the "good life" and indirectly by showing us how others live, providing comparisons that affect self-perceptions and assessment.

Fourth, since we live "the good life" in an environment—neighborhoods and cities for most of us—perceptions of QOL in that environment can be affected in the same manner, through personal observations but also interpersonal, mediated, and mass communication channels and patterns that alert us to opportunities and threats as well as providing us with comparisons of how others live. Witness the numerous top 10 places to live, to visit, or to retire, the best places to work, the best places for singles, the best places for new businesses, etc.

In mass communication, several theoretical traditions and literatures are pertinent. Work in the knowledge gap tradition (e.g., Tichenor, Donohue, & Olien, 1971) reveals that that media use is predictive of community ties and knowledge holding, as newspaper use can increase social capital within communities (e.g., Jeffres, Lee, Neuendorf, & Atkin, 2007). Gerbner and colleagues' (1982, 1986) cultivation theory as well as incidental learning from media use are relevant, suggesting that media cultivate images of our environment, good or bad, and these images become part of our assessment and comparative processes. Agenda setting theory (McCombs & Shaw, 1993), which says the media may not tell us what to think but are quite successful in telling us what to think about, is also applicable; media reports about problems in their surveillance function tell people something negative about QOL in their communities or neighborhoods, and the reverse also is true.

It is the last two of the four areas in which the authors of this paper have worked for much of the past three decades. Here we synthesize those contributions for the significance of communication influences on perceptions of QOL.

METHOD

From 1981 to 2010, we conducted more than two dozen studies in which key QOL measures were included. Most were conducted in the same Midwest metropolitan area; the majority surveyed the entire metropolitan area but several focused on selected city neighborhoods or suburbs. Two companion surveys (to the metro surveys) with QOL items focused on influentials or elites identified through position and domain in the metro area. In addition, two surveys employed national samples, three surveys focused on university students, and five surveys used a panel of ethnics. All used probability samples. The majority of these surveys were conducted by telephone using a computer-aided telephone interviewing (CATI) system, while one used in-person interviews, and the most recent surveys of more specialized populations were conducted using SurveyMonkey (see Table 6.1).

Table 6.1. Summary of Survey Data Sets (1981–2010).

Year	Study Population	Data Collection Method	Sample Size
1981, Spring	Three center city neighborhoods	In-person interviews	161
1982, Spring	Metro Poll & Elite Survey	Telephone & In-person interviews	463/35
1986, Spring	Metro Poll	Telephone	261

1988, Spring	Metro Poll	Telephone	344
1992-1993, Winter	Metro Poll	Telephone, CATI	331
1993, Spring	Metro Poll & Elite Survey	Telephone, CATI	320/25
1993, Fall	Metro Poll	Telephone, CATI	302
1995, Spring	Metro Poll	Telephone, CATI	313
1996, Spring	Metro Poll	Telephone, CATI	377
1999, Fall	6 City, 6 suburb neighborhoods	Telephone, CATI	321
2000, Spring	Metro Poll	Telephone, CATI	351
2000, Fall	Metro Poll	Telephone, CATI	505
2000, Fall	National survey	Internet, commercial data base	2,171
2001, Summer	Metro Poll	Telephone, CATI	305
2001, Fall	Metro Poll	Telephone, CATI	484
2003, Fall	Metro Poll	Telephone, CATI	312
2004, Spring	Metro poll	Telephone, CATI	290
2005, Spring	Metro Poll	Telephone, CATI	142
2006, Fall	Metro Poll	Telephone, CATI	267
2005-2006	National Survey	Telephone, CATI	477
2009, three time periods	Three first ring suburbs	Telephone, CATI	550
2010, Fall	Editors of community newspapers in U.S.	SurveyMonkey	527
1976, 1980, 1984, 1988, 1992	Survey of ethnic panel	Mail and telephone	768/392/363/ 111/157
1999, Summer	Student body	Telephone, CATI	465
2000, Summer	Student body	Telephone, CATI	323
2002, Fall	Student body	Telephone, CATI	505

Note: For a detailed reference list of these surveys and results of additional analyses, please contact Leo W. Jeffres, School of Communication, Cleveland State University, Cleveland, OH 44115.

The key global measure of QOL perceptions in the community was obtained in almost all surveys. This stream of research provides an opportunity to examine with more generalizability than the typical single-shot study. Over the years, our measures of QOL perceptions have evolved to include at various times one or more of the following, using 0–10 scales, 0 = *worst place to live*, 5 = *neutral*, 10 = *the best place to live* (except the last measure): (1) perceptions of QOL in the metropolitan area; (2) perceptions of QOL in the community/neighborhood; (3) perceived quality of specific domains of community life, such as housing, arts/culture, government services; (4) perceptions of life satisfaction; (5) perceptions of satisfaction with specific domains of one's life such as family and work; (6) perceptions of how things are going in the country; (7) for students, attitudes toward the university; (8) for newspaper editors, perceptions of QOL in the community served by their newspaper; and (9) for the ethnic panel, their ethnic identification using a different multi-item scale. The actual measures chosen were used in the classic QOL studies (e.g., Andrews, 1986; Andrews & Withey, 1974; Campbell, 1981; and Campbell et al., 1976) and have demonstrated reliability across different contexts and populations (also see Bowling, 1997; Samli, 1987; Schumacher, 1989). The measures also tap the two dimensions found in a global sense of well-being—personal and societal well-being (Shen & Lai, 1998).

Although the original agenda beginning in 1981 called for examining the relationship between mass media exposure and QOL perceptions, this evolved as the communication environment changed and the studies drew on different theories and research traditions. Thus, our measures of communication evolved to include at various times one or more of the following: hours of watching television; hours of listening to the radio; number of days a week reading a newspaper; number of magazines read regularly; number of books read in past six months; number of videos watched, rented, purchased in past month; number of times went out to see films in a theater in the past month; access to Internet/Frequency one goes on the Internet; frequency one visits media websites; frequency one visits chat rooms; frequency one uses email to send/receive messages; frequency one talks with others about items in the media; neighborhood communication patterns; frequency one reads community/neighborhood newspaper; uses and gratifications for media use; and for the panel of ethnics, their use of ethnic media. Media use was tapped with the usual measures, as were the sociodemographic categories (gender, education, household income, marital status).

RESULTS

QOL in Metro Areas and Communities/Neighborhoods

People's assessments of QOL in their neighborhoods, communities, and the larger metro area where many reside reflect their social economic conditions. As expected, people in wealthier, less congested, and relatively crime-free areas give higher marks. Using the Gallup survey of metro areas in 1978 as a national indicator of urban QOL, we see similar ratings in the Midwest metro area surveyed (range: 5.3 to 7.1, $M = 6.4$). Higher QOL metro assessments are given in suburban areas (7.5), and the influentials surveyed, presumably living in more affluent areas, gave higher QOL ratings too (7.8, 8.3). In the national surveys, ratings for the community as a whole were higher for two of the three studies (7.6 for both) administered by phone; the commercial panel on the web rated their communities lower (6.0).

When asked to rate the specific neighborhood in which they lived, respondents in the metro surveys gave higher QOL assessments, with an average of 7.5 and a narrow range (7.3 to 7.7); results were lower for neighborhood assessments in central city areas (5.9) and higher in the suburbs (7.8), with national studies also high (7.7) and elites highest of all (8.7). In the national survey of non-daily newspaper editors and publishers, the QOL assessment given for the communities they serve was 7.6, similar to the other national ratings. Thus, the subjective assessments likely match the economic conditions creating the environments assessed. Figure 6.1 charts the metro and neighborhood QOL ratings for non-elites across the nearly 30 years spanned by the studies. Clearly, QOL ratings have held steady or possibly increased over the period, with neighborhood ratings exceeding metro ratings systematically over time.

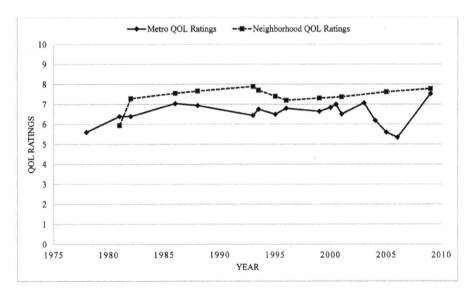

Figure 6.1. QOL Ratings in Metro Areas (1978–2009) and Neighborhoods (1981–2009).

QOL Perceptions and Media Use

Research also shows that people adjust to their circumstances and make comparisons based on what they learn from the media, what they learn from conversations, what they see and experience. Inglehart and Rabier (1986) and Michalos

(1986) propose an aspiration-adjustment model where one's subjective well-being, or perceived QOL, reflects the gap between one's aspirations and one's perceived situation. Since assessments of QOL are affected by the environment, our attention also is drawn to how people perceive that environment and how they make comparisons that ultimately lead to individual assessments, and this leads us to the media and other communication channels that connect people to their environments.

The relationship between more regular exposure to what are now called legacy media, particularly print, is called into question as those media have suffered economic dislocation with the migration of so much advertising to the Internet and social media. It can be argued that at one time anyone who read the daily newspaper or watched local television news regularly was participating in a "community event," since so many other residents, voters, and citizens were engaged at the same time. This also might apply to the local evening news on television, viewed by broad sections of the community. This has been recognized in some uses and gratifications studies. But the media diet has fragmented and members of audiences are increasingly treated not as citizens but consumers and as individuals rather than members of the community. Will this change affect the relationship between regular media exposure—participating in "community media channels"—and community attachment, and perceptions of QOL?

Jeffres et al. (2007) found that newspaper reading was positively correlated not only with both metro and neighborhood QOL assessments but also with community attachment and civic and community values; that study was conducted in 2001. And a similar pattern is found through the years, with relationships found in 12 studies where correlations were statistically significant, approached statistical significance, or emerged with sociodemographic categories controlled. And the relationship persists from 1988 to 2009. None of the other media appears in such a consistent pattern. Time spent watching television is not related, but watching TV news is related to metro QOL in a couple of studies with controls. When time spent listening to the radio is related, it tends to be negative.

As for the correlations between use of traditional media and QOL assessments in the neighborhoods, we saw the same pattern of positive relationships between frequency one reads the newspaper and neighborhood QOL assessments. Only four surveys included measures of TV news viewing and neighborhood QOL assessments, and in three cases, positive relationships were found with sociodemographic categories controlled; the fourth partial correlation approached statistical significance. The rest of the table shows isolated relationships, some negative, some positive.

In summary, it appears that the traditional newspaper has been a positive force for community assessments, despite its traditional function of acting as a watchdog and providing negative news. In the 2010 survey of community newspaper editors and publishers from across the country, respondents' perceptions of QOL available in the community their papers served also was correlated with perceptions that the community has a high consensus on issues (r =.11, p <.01), that the public is interested and not apathetic (r =.33, p <.001), that the community has many local clubs and organizations—a measure of social capital (r =.32, p <.001), that there is a strong sense of community (r =.27, p <.001), and that there is strong community attachment to the newspaper (r =.27, p <.001). Controlling for perceptions that the community is diverse, has a strong economy, and low unemployment did not reduce the relationships.

As for correlations between other communication variables and assessments of the metro quality of life, measures of Internet use were not included until 2000 and relationships began appearing the next year, when visiting media websites was positively related to Metro QOL assessments. Frequency one goes on the web was positively correlated with the Metro QOL assessment once, but negatively correlated in the national survey in 2005–2006 with demographics controlled. Email use was negatively correlated once and a technology use index (summary across a variety of items) was positively related. There was no relationship with visiting chat rooms, a popular phenomenon at that time. Two items tapped communication about current events or politics and both were positively correlated with metro QOL assessments, with sociodemographic categories controlled. Measures of neighborhood/community newspaper readership and a neighborhood communication index tapping interaction with neighbors also tended to be related to metro QOL assessments, as was a communication index crossing media and talking with neighbors in the earlier studies.

As for correlations between other communication variables and assessments of neighborhood QOL, none of the web measures is consistently related to neighborhood QOL assessments, with a couple of negative and one positive relationship. However, the interpersonal communication measures are consistently related to assessments of the neighborhood QOL. This finding of "local" communication being most important is also underlined in the national survey in 2005–2006. Respondents were asked where they would go in their community if they wanted to have a conversation with a friend, a measure of what is referred to as "third places," not work and not home. Oldenburg (1989) defined third places as the "great, good places" that foster community and communication among people outside of home and work, the first two places of daily life. Analysis of this data shows that QOL perceptions were related to the availability of "third places" (Jeffres, Bracken, Jian, & Casey, 2009). So, indirectly, when the environment provides opportunities for personal conversations, people's perceived QOL is higher.

Multivariate Models Predicting QOL Perceptions

Several datasets were analyzed to examine multivariate models predicting QOL perceptions (see Table 6.2). Each study differs in the variables available but all included sociodemographic categories, and some macro-level factors representing the communities in which people live (e.g., population, whether live in urban areas). Research over several decades testifies to the relationship between people's QOL and their personal characteristics and circumstances. Thus, we treated gender, education, income, marital status, and age as influences on people's objective indicators and subjective reports of their QOL—perceived metro and neighborhood QOL, reports of happiness or satisfaction with life overall.

Table 6.2. Multivariate Analyses of Influences Explaining the Good Life/QOL Perceptions.

Year/Dataset	Blocks of Variables	Sig. of Blocks	Significant Betas	Predicting
2009 Survey of three suburbs N = 560	*Social categories Pol. & community variables* (community attachment, no. orgs, community activity, community ties, pol. activity, pol. efficacy) *Communication variables* (media use, freq web, neigh com. Index)	Social categories (*F* ch=5.3 (553) *p* < .001) Pol. & community variables (*F* ch=30.6 (547) *p* < .001) and Com. Variables (*F* ch=2.2 (540) *p* < .04	Age (*β*= .21*) White eth (*β*=-.08#) No. orgs. (*β*=-.12*) Community attachment (*β*=.44*) Pol. efficacy (*β*=.23*) Freq email (*β*=-.14#)	Community QOL *F*=12.8 (540) *p*<.001
	Same blocks	Social categories (*F* ch=3.8 (553) *p* < .001) and Pol. & community vars. (*F* ch=30.1 (547) *p* < .001) explain sig. variance	Age (*β*=.15*) Married (*β*=.14*) Community attachment (*β*=.44*) Pol. efficacy (*β*=.18*) Neigh IP com. (*β*=.12*)	Neigh. QOL *F*=11.5 (540) *p*<.001

Year/Dataset	Blocks of Variables	Sig. of Blocks	Significant Betas	Predicting
2005-2006 National survey $N = 477$	*Social categories* *Community level variables* (diversity, third places avlb.) *Pol. and community vars.* (political efficacy, pol. activity, pol. participation, neighboring, neigh ties, neigh attachment, no. orgs.) *Communication variables* (media use, tech index, web use, com. climate, neigh. IP com., pol. com.)	Social categories (F ch=11.4 (475) $p < .001$) Pol. & community variables (F ch=21.7 (461) $p < .001$) Communication variables (F ch=1.6 (451) $p < .095$).	Age (β=.24*) Income (β=.11*) White eth (β=.12*) Education (β=.08#) No. orgs (β=-.10*) Pol. efficacy (β=.12*) Neigh attachment (β=.47*) Tech index (β=-.14*)	Community QOL F=8.9 (451) p<.001
	Same four blocks	Social categories (F ch=10.5 (475) $p < .001$) Pol. & community variables (F ch=18.3 (461) $p < .001$)	Age (β=.22*) Income (β=.10#) White eth (β=.09*) Education (β=.12*) County pop. (β=-.09#) Neigh attachment (β=.43*)	Neigh. QOL F=7.7 (451) p<.001
2006 Fall Metro Area $N = 267$	*Social categories* *Political behaviors* (pol. activity, pol. efficacy, civic engagement, org. membership) *Com. variables* (media use vars, org. com., pol. com., neigh. com. pattern	Only com. block explains additional variance (F Ch= 2.0*	Civic activity (β=.16*) Reading mags (β=.21*) Reading paper (β=-.12#)	Metro QOL F=1.71 (248), p<.04

Year/Dataset	Blocks of Variables	Sig. of Blocks	Significant Betas	Predicting
2005 Spring Metro area N = 144	*Social categories* *Neigh. variables* (org. involvement, use facilities, neighboring, attachment, ties) *Com.variables* (media use vars. , neighborhood com.)	None of the blocks of variables explain statistically sig. variance	Education (β=.17*)	Metro QOL F=.67 (123), n.s.
	Same 3 blocks of variables	2 blocks explain statistically sig. variance Social categories (F ch.=.15* (117); Com. Variables (F Ch.=2.0* (103)	Age (β=.25*) Reading mags (β=.24*) Reading paper (β=-.18*)	Neigh. QOL F=2.47 (103) p<.01
	Same 3 blocks of variables	Only block of Com. variables approaches significance (F ch=1.8# (124)	Gender (β=.23*) White eth (β=.17#) Freq. Internet (β=-.21*)	Feeling happy F=1.5 (124) p<.08
2004 Spring Metro area N = 302	*Social categories* *Political & community behaviors* (efficacy, interpersonal trust, community attachment, pol. involvement, public affairs knowledge) *Com. variables* (media use, neigh. Com., attention to public affairs news)	2 blocks explain statistically sign. Variance Social categories (F ch=5.4*(299) Pol. behaviors (F ch=28* (293) Com. Variables (F ch= 1.4 n.s.(282)	Education (β=.23*) Gender (β=.12*) Institutional confidence (β=.09*) IP trust (β=-.09#) Community attachment (β=.53*) Read paper (β=-.09#) Watch videos (β=.12*)	Metro QOL F=10.3 (282) p<.001

Year/Dataset	Blocks of Variables	Sig. of Blocks	Significant Betas	Predicting
2003 Fall Metro area $N = 314$	*Social categories* *Com. Variables* focusing on U&G, presence, leisure activities (popular culture knowledge index, TV uses and grats index, parasocial U&G Index, enjoyment of leisure activities index, media use)	Only block of social categories approaches significance (F ch=2.1 (307) $p < .06$	Age ($\beta=.13^*$) Pop. culture knowledge ($\beta=-.12^*$) Parasocial U&G index ($\beta=.16^*$) Watching videos ($\beta=-.11\#$)	Metro QOL $F=1.5$ (293) $p<.07$
	Same 2 blocks of variables	Neither block sig.	Hours visit web-sites weekly ($\beta=.10\#$)	How happy am overall $F=.95$ (293) n.s.
2001 Summer Metro area $N = 305$	*Social categories* *Leisure and Values* (attending events index, factor scores for importance of different values) *Com. variables* (media use, freq. Internet)	Social cate-gories block approaches sig. (F ch=2.0 (298) $p < .08$) Leisure & values sig. (F ch=4.1 (289) $p < .001$. Com. Vars not sig.	White eth ($\beta=.15^*$) Events index ($\beta=.16^*$) Politics & com-munity values ($\beta=.18^*$) Comfort & friends values impt. ($\beta=.12^*$) Individualism values impt. ($\beta=.11^*$)	Metro QOL $F=2.2$ (278) $p<.001$
	Same three blocks	2 blocks explain statistically sign. Variance Social catego-ries (F ch=6.2 (298) $p < .001$ Leisure & Values (F ch=4.3 (289) $p < .001$	Age ($\beta=.15^*$) White eth ($\beta=22^*$) Income ($\beta=.11\#$) Events Index ($\beta=.13^*$) Comfort & friends values impt. ($\beta=.24^*$)	Neigh. QOL $F=3.8$ (278) $p<.001$

Year/Dataset	Blocks of Variables	Sig. of Blocks	Significant Betas	Predicting
		Com. Vars (*F* ch=1.6 (278) *p* < .10	Competition & modernity values impt. (*β*=.10#)	
			No. books read (*β*=.12*) Read paper (*β*=-.11#)	
2000 Fall Metro area *N* = 505	*Social categories* *Political variables* (watching debates, pol. knowledge, party ID) *Com. variables* (media use, freq. talk politics)	2 blocks explain statistically sig. Variance Social categories (*F* ch=4.2 (498) *p* < .01 Pol. variables (*F* ch=3.0 (495) *p* < .03	Age (*β*=.19*) Income (*β*=.08#) Watching debates (*β*=.12*) Reading paper (*β*=.11*) Listen radio (*β*=-08#)	Metro QOL *F*=2.2 (481) *p*<.001
1999 Fall Suburbs, city neigh- borhoods *N* = 389	*Social categories* *Political and commu- nity variables* (faith in civic activities, neighborhood activities, neigh. attachment) *Com. Variables* (media use, Internet access & use, neigh. com, neigh. papers)	Only pol. and community variables block explains sig. variance (*F* ch=2.6 (381) *p* < .05).	Age (*β*=.13*) Neigh. attach- ment (*β*=.16*) Listen radio (*β*=-.10*) Freq read paper (*β*=-.11*)	Metro QOL *F*=1.3 (368) n.s.
	Same 3 blocks	2 blocks explain statistically sig. variance Social catego- ries (*F* ch=5.5 (384) *p* < .001)	Age (*β*=.18*) Gender (*β*=.12*) Income (*β*=.10#) Neigh. attach- ment (*β*=.52*)	Neigh. QOL *F*= 8.2 (368)) *p*<.001

Year/Dataset	Blocks of Variables	Sig. of Blocks	Significant Betas	Predicting
	Political & community variables	Faith civic act ($\beta=-.09\#$)		
		(F ch=40.1 (381) $p < .001$	Neigh. Com ($\beta=-.13^*$)	

Note: $^* p < .05$; $\# p < .10$.

In addition, we factored in the nature of the communities as potential influences when those data were available. And, in some studies, additional influences were measured that elaborated on personal interests and individual behaviors; these too were included. That takes us to the focus of this chapter—communication, which not only links people to their environment by providing information (family, friends, community) but also can be seen as more instrumental affectively—helping people feel "good" about their relationships, for example. Using regression analyses, we examined the available data to see how these sets of influences explain variance in people's reports of the "good life" as represented by QOL measures.

QOL perceptions and intent to stay in communities were examined in multivariate analysis using the 1981 neighborhood dataset and 1982 metro survey dataset (Jeffres, Dobos, & Sweeney, 1983). In the neighborhood model, demographics predicted to interpersonal communication and newspaper reading, and structural links (ties to area); these then impacted beliefs about the neighborhood problems and assets, which impacted neighborhood QOL assessments and how happy residents were with their neighborhood. And this predicted intentions to stay in the neighborhood. In the metro model, demographics were exogenous variables, with newspaper readership and viewing TV news impacting perceptions of problems and assets of the metro area, which then impacted metro QOL assessments. Metro QOL assessments, neighborhood QOL assessments, and perceptions of how others would rate the area (subjective norms), then predicted intent to stay in the metro area.

The 1982, 1986, and 1988 studies also measured how valued opportunities for leisure in the area were and assessed on 0–10 scales the quality of spectator and cultural events and outdoor recreation. For each data set, path models were constructed, with media use and values affecting metro QOL. In a second study, media use and leisure values predicted assessments of leisure opportunities, which affected metro QOL. And in the third data set, a communication index (of TV news viewing, newspaper reading, and interpersonal neighborhood communication patterns) affected leisure images of the area but not metro QOL. In all three models demographics were exogenous variables (Jeffres & Dobos, 1993).

The 1982, 1986, and 1988 datasets also were analyzed to test a bottom-up model predicting one's satisfaction with life, with the same demographic exogenous variables, communication variables and comparison processes (comparisons with other metro areas), and assessments of personal domains (family, work, leisure); results showed that domain assessments did not explain additional variance beyond that accounted for by the demographics and communication variables, meaning the data failed to support the bottom-up view (Jeffres & Dobos, 1995).

Applying the cultivation model where mass media impact metro QOL assessments through cumulative exposure, Jeffres, Neuendorf, and Atkin (2000) examined six survey data sets, 1992 to 1999, creating indices across standardized media use variables. While the media as a whole failed to cultivate a consistent picture of the metro QOL, there was some tendency for print media to be positively related and broadcast media (radio, TV, film, video) to be negatively related to QOL assessments. However, this pattern reflected by bivariate and partial correlations should be accompanied by a fuller context.

Table 6.2 summarizes multivariate analyses explaining QOL perceptions, using hierarchical regression entering sociodemographic categories in the first block, other community, political, or civic engagement measures featured in the particular study as a second block, and then communication variables in the third block. Separate analyses were conducted predicting metro QOL assessments, neighborhood QOL assessments, and measures of personal happiness or satisfaction.

Looking at regressions on metro and neighborhood QOL assessments across the diverse set of studies, we see that sociodemographic categories perform as expected when statistically significant—older, married, and higher status people rated higher QOL in their communities. While the second block includes different variables from study to study, measures of civic engagement, political efficacy, civic values, and community attachment also predicted a higher metro and neighborhood QOL when related at all (interestingly, belonging to organizations—often seen as social capital—is negatively related in the two studies where it's measured). Finally, the third block of communication variables shows a mixed picture, with one medium standing out: Reading a newspaper more regularly is a negative predictor most of the time. Twice time spent listening to the radio is a negative predictor, and measures of newer technologies also seem to cast a negative influence (email use and access to more technologies in 2005, 2006, and 2009) as do neighborhood communication measures. Regressions on reports of how things are going in one's personal life or happiness show few significant relationships.

Earlier analyses using one of the data sets showed that exposure to community media is associated with community and neighborhood attachment, which, in the multivariate analyses appears to wipe out independent influence by media and other communication variables. To investigate this across the various studies, we examined relationships between media use and other communication variables, and the community, civic engagement, and political variables entered in the regression analyses. The pattern of results across the studies confirms the strong relationship between newspaper readership and community attachment, neighborhood ties, political activity, and organizational membership. Similarly, reading the paper leads to higher knowledge scores for public and political affairs, popular culture, and the broader ethnic-religious cultures. Time spent watching television is not associated with community attachment but there is some displacement with lower community activity, neighboring, and political engagement.

While the pattern is mixed for watching TV news, overall there's a positive relationship between viewing and community attachment, with isolated statistically significant partial correlations with other variables in the mix. Listening to the radio is unimportant here, but frequency of Internet use appears to be positively associated with both community attachment, contrary to many fears, as well as community and political activities. Some of this Internet use appears to be visits to media websites, a variable included in several studies; controlling for sociodemographic categories, visiting media websites was correlated with political participation (partial r =.15, p <.01) and organizational membership (partial r =.11, p <.02) in the national 2005–2006 survey, with community attachment (partial r =.20, p <.01) in the 2009 survey, with how happy one was overall (partial r =.16, p <.01) in the 2003 fall survey, and with community involvement (partial r =.13, p <.03) in the 2001 summer survey. The neighborhood communication pattern, measured generally as an index of interaction with neighbors and others in the community, is strongly related to not only community attachment but almost all the variables tapping political and community activities and political efficacy. Similarly, though only appearing in two of the studies, a strong political discussion network is positively related to community attachment, neighboring, efficacy, and participating in political activities.

This suggests that the role performed by media and interpersonal communication patterns is to influence people's relationships with and involvement in their communities rather than perceptions of overall levels of QOL. In other words, reading the paper and talking with neighbors makes one feel as if they're part of a community — but also provide the news and information — leading to a more critical analysis of the QOL offered by their environment.

Returning to the data sets with neighborhood involvement and attachment measures, we reran the regressions predicting metro and neighborhood QOL. For the 2009 survey, excluding the neighborhood attachment and involvement measures, neighborhood interpersonal communication emerges as a significant positive predictor of community (β =.20, p <.001) and neighborhood (β =.32, p <.001) QOL assessments. Similarly, the same neighborhood interpersonal communication measure emerges as a significant positive predictor for community (β =.12, p <.05) and neighborhood (β =.24, p <.001) QOL in the 2005–2006 national survey. In the 2004 spring survey analysis, the negative betas for newspaper readership and watching videos drop out and positive betas appear for time spent watching TV (β =.18, p <.01) and frequency of Internet use (β =.19, p <.01) in predicting metro QOL. And in the 1993 survey of suburbs and city neighborhoods, there's no similar change in predicting Metro QOL, but the negative beta for neighborhood communication in predicting neighborhood QOL drops out. Thus, there's some support for the notion that communication variables are closely linked to community attachment and involvement, which then impact QOL assessments.

To tease out relationships among these variables, we constructed a path model using the key communication variables using the spring 2000 dataset. The model incorporates two demographic variables, public affairs interests, newspaper readership, knowledge of the metro area, neighborhood and metro QOL assessments; the variables were drawn from the larger array of measures where their bivariate relationships were significant. As Figure 6.2 shows, newspaper readership predicts knowledge of the metro area, which then predicts metro QOL assessment. A second path suggests that communication variables influence other domains, which then impact the overall QOL assessments. In six studies, respondents were asked to assess their QOL in personal domains (e.g., work, family, friends) or community domains (e.g., housing, schools, law). While the former should predict the global satisfaction with life, the latter should predict assessments of the community quality of life. Jeffres and Dobos (1995) found some support for this proposition.

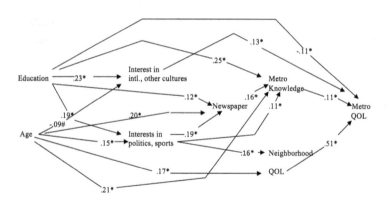

Figure 6.2. A Path Model of Relationships Predicting Assessments of the Quality of Life in a Community (Metro Poll).

The 1986 study included assessments of the quality of specific metro domains, including public schools, housing, justice and law enforcement, government services, public transportation, employment opportunities, cultural events, and outdoor recreation, and the 1988 study included assessments of the quality of personal domains, including friends, spouse/family, job, groups/clubs, church, hobbies and interests. Mass communication variables were related to six assessments of the nine environmental domains: Watching local television news positively related to assessments of the culture and arts domain in 1986 (r =.20, p <.05), and newspaper readership was correlated with assessments of culture and arts (1986, r =.14, p <.05), housing (1986; r =.16, p <.05), outdoor recreation (1986, r =.13, p <.05) and two of three neighborhood QOL assessments (1982, r =.11, p <.05; 1988 r =.15, p <.05). Interpersonal communication measures also were related to assessments of environmental domains: in 1986 frequency talking about local problems correlated with government services (r = -.16, p <.05), and jobs-employment (r =.13, p <.05); frequency of neighborhood communication was correlated with neighborhood QOL in 1988 (r =.21, p <.05).

In the more recent data sets, we find some discernible patterns involving media variables. Newspaper readership is positively related with family life satisfaction in the 1993–1999 studies, and the relationships persist with sociodemographic categories controlled. Listening to the radio is negatively associated with family life satisfaction, and there are a few other isolated correlations. Satisfaction with work appears in a mixed pattern of positive and negative relationships with media variables during this period.

Shifting to metro domains, examined in 1992 and 2009 studies, we see differentiation so that, for example, reading the paper and watching more TV were positively associated with assessments of the police in 1992 but negatively associated with assessments of public transit and going to see films more often was associated with lower assessments of both. While newspaper readership drops out of relationships in 2009, watching more TV news, using email more often and frequent Internet use appear in negative relationships with assessments of schools, police, and beliefs about how people treat each other.

As for relationships between media use and reports of happiness, satisfaction with personal life, sometimes reading the newspaper is related to reports of satisfaction with how things are going in one's personal life (1993, 1999), but at other times there's a negative relationship with how happy one feels (2004). Listening to the radio more often also appears in negative relationships (1993, 1999), while web access (1999), frequency of web use (2004) and visiting media websites (2003) are positively related. Uses associated with particular television formats (talk shows and sitcoms) is positively related to reports of happiness, while using TV to keep in touch is negatively related.

QOL Perceptions for Specialized Populations

Beginning in 1976 and continuing to 1992 surveys were conducted in presidential election years using a diverse panel of ethnics. Ethnic identification, as measured through a series of items tapping the importance of one's ethnic culture, is a useful indicator of the "good life" for these respondents. Data show a strong pattern of relationships between ethnic media use and ethnic identification across all waves, and cross-legged correlation analysis shows that the strength of ethnic media use in 1976 was positively associated with feeling closer to one's ethnic heritage in 1992, while the reverse relationship was not true (Jeffres, 1999, 2000).

Students at a Midwestern university also were surveyed on the quality of the institution they attended in 1999, 2000, and 2002. Student communication patterns at the university were strongly correlated with both attitudes and assessments. And, consistent with the metro data, readership of campus publications and newspapers in general also was correlated with attitudes toward the institution. One survey asked students to rate QOL in the metro area, which was positively related to newspaper readership and negatively related to time spent watching television and frequency of Internet use.

SUMMARY AND DISCUSSION

Research into the "good life" has recognized that people's QOL assessments may be affected by their assessment of the larger environment and its impact on them. This chapter reports on QOL assessments examined by the authors over 30 years, drawing on a variety of metropolitan, community/neighborhood, and national data sets that included diverse measures of communication.

In summary, we learned that: (1) The larger environment matters, with higher QOL assessments in wealthier areas and communities. (2) Sociodemographic categories also matter, with higher QOL ratings given by elites than the general public. Higher QOL is associated with white ethnicity, being married, being older, higher education, and higher incomes. (3) Other individual differences also matter, even with demographics controlled, with some studies showing most positive QOL assessments related to political efficacy, community and neighborhood attachment, civic activity, going to local events, confidence in institutions, and values (valuing community, politics, friends, individualism, competition, and modernity). (4) People think that QOL in their more immediate neighborhood is higher than in the larger community.

Against these results, we focused on the relationship between communication variables, particularly media use, and QOL assessments, learning that: (1) Newspaper reading was positively correlated with both metro and neighborhood QOL assessments over the years (controlling for sociodemographic categories), with none of the other media appearing in a similar consistent pattern except watching TV news, which was positively related to metro QOL assessments with controls in some studies. The prominent role played by newspapers here confirms national studies demonstrating the medium's prominence in QOL domains, about which a recent Pew study (2011) concluded that "More than half of regular newspaper readers (56%) say that if the local newspaper they read most often no longer published—either in print or online—it would hurt the civic life of the community a lot" (p. 1). (2) Examining assessments of domains, newspaper reading was positively related to assessments of family life in several studies and with some metro domains in a couple studies in the 1990s, while watching TV news was positively related and both email and Internet use negatively related to some metro domains (schools, police, how people treat each other) in 2009. (3) Reports of personal happiness and satisfaction with personal life were positively related to newspaper reading in three studies and negatively related in one study from 1993 to 2005. Reports of happiness were not related to any media use and to only a couple content preferences and uses and gratifications items in a 2003 study. (4) Studies of specialized populations show significant relationships between student communication patterns and attitudes/assessments regarding the institution, and between ethnic media use and ethnic identification in a study of different ethnic groups from 1976 to 1992.

Results show that the most "local" channels of and opportunities for communication—talking with neighbors, reading neighborhood papers, and the availability of third places—enhance perceptions of the community QOL. Thus, we found that (1) Interpersonal communication variables (interaction with neighbors) are related to neighborhood QOL assessments; (2) Reading neighborhood/community papers is related to neighborhood and metro QOL assessments; and (3) Availability of "third places" is positively related to community QOL assessments.

However, the direct influence of communication variables largely drops out in multivariate analyses. Controlling for sociodemographic categories and other individual differences, the influence of newspaper reading, media use, and neighborhood communication patterns in general drops out or turns negative as influences on community QOL assessments. This result stems from a strong relationship between the communication variables and community attachment, neighborhood ties, political activity, and organizational membership.

Thus, the function of media and interpersonal communication patterns is to influence people's relationship with and involvement in their communities rather than QOL perceptions. So reading newspapers and talking with neighbors make people feel as if they are part of a community but also provide news and information leading to a more critical analysis of life offered by the environment. The results are consistent with media effects theories such as agenda setting and cultivation.

The media environment has changed dramatically since the authors began collecting these data sets, with the media increasingly fragmenting. The legacy media—newspapers and TV news—which report on the geographic communities in which people live were the most important influences on QOL assessments over this period. Readership and watching the local TV news also are correlated with community attachment and community ties that are related to not only QOL assessments but also a host of measures of civic engagement. Daily newspaper circulations have declined as advertising losses have shaken its financial base. Television stations also have suffered as advertising online has grown. However, there is less evidence that the web has had much impact on non-daily newspapers (Jeffres & Kumar, 2011a, 2011b), which continue to thrive by serving communities with local news.

We need to ask whether newspapers and TV news operations can continue their influence in attaching people to their communities, and whether online audiences will experience the same "sense of community" produced by legacy media. During this period, there was some positive evidence linking online activity to community attachment and civic engagement, but this will need to be updated with the advent of social media such as Facebook, which places the individual at the center of a network rather than linking the individual to a community-based enterprise. While i-neighbors and other web-based efforts attempt to connect people within geographic communities, anecdotal evidence suggests that they speak to the committed rather than strengthening ties of those with weaker attachments.

In a sense, when one is on the Internet, one is located "in no particular place." To its great and considerable advantage, when you are on the Internet you are "everywhere," or "anywhere," but perhaps also "nowhere." And the impact of legacy media may in part stem from the fact that using them is a "communal" experience as much as it is a consumer experience. Unfortunately, this set of studies evolved over several decades and did not include the same variables consistently, so studies that update in a comprehensive way will be needed to tease out how media and interpersonal communication patterns in the community continue to affect our assessments of the "good life."

It will be important, in later work, to continue this research as our media environment is transformed by emerging wireless telematics platforms. For instance, according to one source, 2010 was the first year in which more people reported receiving more of their news from online rather than print sources (Pew, 2011). Moves to enhance interactivity and dynamic content may build community among readers (Imfeld & Scott, 2005), but that may be to the detriment of "geographic communities" in which people live as they gravitate toward like-minded folks regardless of their "community" rather than neighbors or fellow "citizens." We might expect to see the role of interpersonal influences grow as emerging "intermass" forums (e.g., Facebook) are driven by interpersonal applications (e.g., Hunt, Atkin, & Kowal, 2013). Further research can profitably assess these QOL dynamics in more diverse physical and media environments.

Surveys at the national level have asked Americans for years whether they think the country is headed in the wrong direction, or whether they are satisfied or dissatisfied with how things are going in the country. Such items have been taken as "global" attitudes toward actors in the public arena, and they have been significant predictors of voting sentiment. Similarly, when people are asked to assess the QOL in their communities, they are asked to provide a measure of how they "feel" about their community. Additional measures that ask how "happy" or "satisfied" one is to achieve a more personal, perhaps temporary, measure of the status of one's life. A couple of the studies here broke down assessments into personal domains (e.g., family, work) and public domains (e.g., schools, public services). The interplay among these domains deserves attention because people experience the full range of domains and live in a neighborhood, a community, and a country. Communication not only connects people but it also links them to each of these domains and levels, raising the issue of how a changing communication environment will impact our assessments of the "good life" as it is interpreted within and across contexts. Galbraith a generation ago said that the role of advertising in the capitalist system was to show consumers there was always something new, something more, so they could never be satisfied and had a reason for being productive. In this scenario, the "good life" is never achieved, only a goal sought after. Now that there's a common set of material aspirations around the world (Mahbubani, 2013), people are more likely to accept a consumption definition of what constitutes the "good life." But material values do not trump all, as religious differences, ideological and philosophical disagreements about the environment, and national political aspirations can change people's assessments of the "good life" in short order. And at the individual level, people with difficult personal circumstances and a negative QOL assessment may find compensation in a sport team whose achievements make them (and their community) feel like winners. Some assessments of the "good life" may be fleeting, others amazingly durable, but it is certain that communication processes are involved in the change.

REFERENCES

Andrews, F. M. (1986). *Research on the quality of life.* Ann Arbor, MI: Survey Research Center, Institute for Social Research, University of Michigan.

Andrews, F. M., & Withey, S. (1974). Developing measures of perceived life quality: Results from several national surveys. *Social Indicators Research, 1*(1), 1–26.

Belk, R. W., & Pollay, R. W. (1985). Images of ourselves: The good life in twentieth century advertising. *Journal of Consumer Research, 11*(4), 887–897.

Bowling, A. (1997). *Measuring health: A review of quality of life measurement scales.* Philadelphia: Open University Press.

Campbell, A. (1981). *The sense of well-being in America: Recent patterns and trends.* New York: McGraw-Hill.

Campbell, A., Converse, P., & Rodgers, W. (1976). *The quality of American life: Perceptions, evaluations, and satisfactions.* New York: Russell Sage.

Carbonell, X., Guardiola, E., Beranuy, M., & Bellés, A. (2009). A bibliometric analysis of the scientific literature on Internet, video games, and cell phone addiction. *Journal of the Medical Library Association, 97*(2), 102–107.

Cherry, C. (1971). *World communication: Threat or promise?* New York: Wiley.

Dittmar, H. (2007). The costs of consumer culture and the "cage within": The impact of the material "good life" and "body perfect" ideals on individuals' identity and well being. *Psychological Inquiry, 18*(1), 23–31.

Dye, T. (2010). *Understanding public policy* (13th ed.). New York: Pearson.

Eisenberg, R. (2013, August 9). What 'the end of the suburbs' means for boomers. *Forbes.* Retrieved from http://www.forbes.com/sites/nextavenue/2013/08/09/what-the-end-of-the-suburbs-means-for-boomers/

Gerbner, G. (1982). Introductory comments. In D. Pearl, L. Bouthilet, & J. Lazar (Eds.), *Television and behavior: Ten years of scientific progress and implications for the eighties* (DHHS Publication No. ADM 82–1196, Vol. 2, pp. 332–333). Washington, DC: U.S. Government Printing Office.

Gerbner, G., Gross, L., Morgan, M., & Signorielli, N. (1986). Living with television: The dynamics of the cultivation process. In J. Bryant & D. Zillmann (Eds.), *Perspectives on media effects* (pp. 17–40). Hillsdale, NJ: Erlbaum.

Headley, B., Veenhoven, R., & Wearing, A. (1991). Top-down versus bottom-up theories of subjective well-being. *Social Indicators Research, 24*(1), 81–100.

Hunt, D., Atkin, D., & Kowal, C. (2013). Community attachment affects use of online, interactive features. *Newspaper Research Journal, 34*(2), 64–77.

Imfeld, C., & Scott, G. W. (2005). Under construction: Measures of community building at newspaper web sites. In M. B. Salwen, B. Garrison, & P. D. Driscoll (Eds.), *Online news and the public* (pp. 205–220). Mahwah, NJ: Erlbaum.

Inglehart, R., & Rabier, J. (1986). Aspirations adapt to situations-but why are the Belgians so much happier than the French? In F. M. Andrews (Ed.), *Research on the quality of life* (pp. 1–56). Ann Arbor, MI: Survey Research Center, Institute for Social Research, University of Michigan.

Jeffres, L. W. (1999). The impact of ethnicity and ethnic media on presidential voting patterns. *Journalism & Communication Monographs, 1*, 198–262.

Jeffres, L. W. (2000). Ethnicity and ethnic media use: A panel study. *Communication Research, 27*(4), 496–535.

Jeffres, L. W., Bracken, C., Jian, G., & Casey, M. (2009). The impact of third places on community quality of life. *Applied Research in Quality of Life, 4*(4), 333–345.

Jeffres, L. W., & Dobos, J. (1993). Perceptions of leisure opportunities and the quality of life in a metropolitan area. *Journal of Leisure Research, 25*(2), 203–217.

Jeffres, L. W., & Dobos, J. (1995). Separating people's satisfaction with life and public perceptions of the quality of life in the environment. *Social Indicators Research, 34*(2), 181–211.

Jeffres, L. W., Dobos, J., & Sweeney, M. (1983, November). *Communication and commitment to community.* Paper presented at the annual convention of the Midwest Association for Public Opinion Research, Chicago, IL.

Jeffres, L. W., & Kumar, A. (2011a, July). Survey: Community newspapers have a bright future. *Publishers Auxiliary*, pp. 18–20.

Jeffres, L. W., & Kumar, A. (2011b, August). Study: Community papers increasing online presence. *Publishers Auxiliary*, pp. 1–20.

Jeffres, L. W., Lee, J., Neuendorf, K., & Atkin, D. J. (2007). Newspaper reading, civic values and community social capital. *Newspaper Research Journal, 28*(1), 6–23.

Jeffres, L. W., Neuendorf, K., & Atkin, D. (2000). Media use patterns and public perceptions of the quality of life. *Proceedings of the Second International Conference on Quality of Life in Cities* (Vol. 2, pp. 369–385). Singapore: National University of Singapore.

Jeffres, L. W., Neuendorf, K., & Atkin, D. (2012). Acquiring knowledge from the media in the Internet age. *Communication Quarterly, 60*(1), 59–79.

Lane, R. E. (2000). Diminishing returns to income, companionship—and happiness. *Journal of Happiness Studies, 1*(1), 103–119.

Mahbubani, K. (2013). *The great convergence: Asia, the West, and the logic of one world.* New York: Public Affairs.

Maslow, A. H. (1968). *Toward a psychology of being.* Princeton, NJ: Van Nostrand.

McCombs, M., & Shaw, D. (1993). The evolution of agenda-setting research: Twenty-five years in the marketplace of ideas. *Journal of Communication, 43*(2), 58–67.

Michalos, A. (1986). Job satisfaction, marital satisfaction and the quality of life: A review and a preview. In F. M. Andrews (Ed.), *Research on the quality of life* (pp. 57–83). Ann Arbor, MI: Survey Center, Institute for Social Research, University of Michigan.

Oldenburg, R. (1989). *The great good place.* New York: Marlowe.

Perry, S. D., & Lee, K. C. (2007). Mobile phone text messaging overuse among developing world university students. *Communication: South African Journal for Communication Theory and Research, 33*, 63–79.

Pew Research Center's Project for Excellence in Journalism. (2011, March). *The state of the news media 2011: An annual report on American journalism.* Retrieved from http://stateofthemedia.org/overview-2011/

Samli, A. C. (Ed.). (1987). *Marketing and the quality of life interface.* Westport, CT: Quorum Books.

Schumacher, R., et al. (1989). *World quality of life indicators.* Santa Barbara, CA: ABC-CLIO.

Shen, S., & Lai, Y. (1998). Optimally scaled quality-of-life indicators. *Social Indicators Research, 44*, 225–254.

Smilansky, S. (2012). Life is good. *South African Journal of Philosophy, 31*(1), 69–78.

Spiegel, J. (2012, Dec. 21). Hanging up on cell phone addiction. What to do when an electronic device becomes your best friend. *Psychology Today.* Retrieved from http://www.psychologytoday.com/blog/mind-tapas/201212/hanging-cell-phone-addiction

Tichenor, P., Donohue, G., & Olien, C. (1971). Mass media flow and differential growth in knowledge. *Public Opinion Quarterly, 34*(2), 159–170.

UPI. (2013, Aug. 10). Baby boomers giving up suburban homes for city life. *UPI.* Retrieved from http://www.upi.com/Top_News/US/2013/08/10/Baby-boomers-giving-up-suburban-homes-for-city-life/UPI-25491376166067/

Tuning in versus Zoning out

The Role of Ego Depletion in Selective Exposure to Challenging Media

ALLISON EDEN AND TILO HARTMANN
VU UNIVERSITY AMSTERDAM, THE NETHERLANDS

LEONARD REINECKE
JOHANNES GUTENBURG UNIVERSITY MAINZ, GERMANY

Increasingly, people live in an "option-rich" media environment that provides a plethora of entertainment and information opportunities (Vorderer & Kohring, 2013, p. 188). While these developments have considerably expanded our opportunities for relaxation, personal growth, and self-actualization via media, they also offer individuals the chance to tune out, disengage, and opt for "guilty pleasures" that address short-term hedonic needs versus long-term personal goals (Panek, 2014). The current study examines specific determinants of media choice, which may lead to media use that either challenges or does not challenge the viewer.

We define challenge in terms of the extent to which media require self-regulatory resources. Our rationale is simple: If users have already exhausted their existing self-control resources, they may be less inclined to choose and process media content that requires self-control to process, versus content that does not draw upon their already depleted resources. Over time, these types of media choices may lead to a pattern of hedonically motivated media consumption, instead of that which challenges, engages, or expands the horizon of the user.

Considering media challenge in terms of self-regulatory resources is not without precedent. Hartmann (2013) and Bartsch and Hartmann (2015) have both recently attempted to explain the role of challenge in viewers' entertainment experience via this perspective. Bartsch and Hartmann (2015) found that media content can be differentiated in terms of the types of challenge (cognitive or affective) it may offer viewers, and that the types of challenge presented by content can be linked to motivations to consume this content in an entertainment setting; however, they did not link the preference for challenging content to the resources available to the user at the time of choice, which is the focus of the current research. In this chapter, we posit that the appeal of challenging content may depend on the available self-regulatory resources of the user at the time of choice.

Self-regulatory capacity, or self-control, is the ability to control behaviors and desires in order to obtain a desired goal. It has been extensively researched as a key variable predicting individual success in domains such as education, health, and personal relationships (Baumeister, Vohs, & Tice, 2007), as well as leisure time (Hofmann, Vohs, & Baumeister, 2012). Continued, effortful self-regulation is theorized to drain a limited psychological resource and lead to a state known as "ego depletion," which may then lead to people making short-term, hedonic choices that may not benefit their long-term goals (Gailliot et al., 2007; Muraven & Baumeister, 2000). There is some research to suggest that individuals who are in an ego-depleted state are less able to resist the temptations presented by media. For example, Wagner, Barnes, Lim, and Ferris (2012) showed that increased ego depletion is a strong predictor of "cyberloafing" at the office. However, Panek (2014) and Reinecke, Hartmann, and Eden (2014) found no relationship between ego depletion and duration of media use, suggesting that key moderator variables are missing from the understood relationship between self-control and media indulgence.

Given that ego depletion leads to choices that may appeal in the moment but not have lasting cognitive or physical benefit, we posit that when in front of the screen, ego-depleted people may be more prone to choose media that are considered passive entertainment, or "lean back" media, rather than intellectually or emotionally challenging media content. This would fit with ego depletion research suggesting that users prefer hedonic, immediately satisfying stimuli when they are in a depleted state. Reinecke et al. (2014) offer partial support for this notion, showing that ego depletion was negatively related to selection of challenging television content. Similarly, Panek (2014) found that students' self-control was negatively related to the use of online video and social network services, but not to movie or DVD use. In self-control terms, we would suggest that perhaps different media require different levels of self-regulation by users. Media content that is more challenging, because it demands higher levels of cognitive or emotional self-control, may thus be less appealing for media users whose self-regulatory capacities have been depleted by prior activities.

In our current research, we focus on experimentally testing the role of ego depletion on users' preference to choose challenging media content. We begin by defining ego depletion, then discuss how ego depletion may affect media choice, next how challenge induced by media stimuli may be conceptualized, and finally present two experimental studies investigating the role of ego depletion in media choice.

Self-control and the Depleted Self

Self-control is a crucial form of human execute functioning and refers to "the capacity for altering one's own responses, especially to bring them into line with standards such as ideals, values, morals, and social expectations, and to support the pursuit of long-term goals" (Baumeister et al., 2007, p. 351). Psychological research has conceptualized self-control both in terms of trait-like individual differences and in terms of situational fluctuations in self-regulatory capacity. The strength model of self-control (Baumeister et al., 2007) proposes that the exertion of volition consumes a limited resource. All forms of behavior that demand self-regulation draw on this limited self-control capacity. Preceding acts of self-regulation, thus, deplete the limited self-regulatory capacity and make subsequent self-control more difficult. This temporary impairment of self-control is referred to as ego depletion: "a temporary reduction in the self's capacity or willingness to engage in volitional action caused by prior exercise of volition" (Baumeister, Bratslavsky, Muraven, & Tice, 1998, p. 1253). Ego depletion results from effortful and exhausting prior self-regulation, such as making decisions, complying with social norms, preventing errors in cognitive tasks, persevering against challenges, and in many other tasks encountered in daily life (Baumeister & Heatherton, 1996).

Media researchers have only recently started focusing on self-regulation, and the lack thereof, as a contributor to media selection and consumption processes. For example, a recent time sampling study by Hofmann et al. (2012) suggests a strong link between ego depletion and media use. The results show that the more self-regulation a person engages in during the day, the less able they are to regulate subsequent behaviors. Furthermore, the authors identify media use as the desire least likely to be successfully regulated in the face of conflicting desires. This suggests that media use may present a self-regulatory challenge, and one that is likely to be unsuccessfully met. Wagner et al. (2012) showed that reduced self-control capacity is a strong predictor of personal media use during work hours. In a panel study by Brosius, Rossmann, and Elnain (1999) individuals who felt more stressed after work (perhaps due to self-regulatory failure, although this is not included in the study) spent more time watching entertainment than others. Furthermore, the results of a survey among users of browser games by Reinecke (2009) demonstrate that employees who experience higher levels of work-related fatigue use video games more frequently during working hours than less chronically fatigued individuals. Although Reinecke et al. (2014) did not find a link between ego depletion and the hours spent with entertainment, they did find that ego depletion affected the choice of less challenging media content by viewers, and the appraisals of that media use.

These studies together suggest that media may represent a hedonic temptation that is hard for ego-depleted people to resist or control: Ego-depleted individuals are drawn to pleasant and unchallenging activities (Shiv & Fedorikhin,

1999) and entertaining media content seems to provide a perfect fit with this desire. Entertaining media stimuli do, however, vary considerably with regard to the cognitive and emotional challenges they pose to the media users. While "lowbrow" forms of entertainment seem to perfectly meet the needs of ego-depleted individuals, more challenging media may rather pose an additional challenge to the already low self-regulatory capacities of these users. Instead of helping to restore self-control or at least providing a break from self-regulatory demands, such media stimuli might thus even further deplete the individual's self-control. This may have an impact on how attractive these media are to ego-depleted viewers during selection processes. The next section considers media content in terms of challenge, defined as the amount of self-regulatory capacity demanded by the emotional and cognitive processing of specific media messages.

Cognitive and Emotional Processing of Media Content as a Challenge to Self-control

Selection of entertaining media content has been primarily considered a hedonic regulation mechanism in past literature (e.g., Zillmann & Bryant, 1985), perhaps due to the focus on enjoyment as the ultimate goal of entertainment consumption (Vorderer, Klimmt, & Ritterfeld, 2004). However recent conceptualizations of entertainment suggest that media may offer more than simple hedonic pleasures to the viewer (Oliver & Raney, 2011). For example, Oliver and Raney (2011) suggest that entertainment may also offer the opportunity for viewers to learn about their lives or search for meaning. They found that motivations for viewing movies could be split into hedonic motivations (to have fun, laugh, and enjoy silly entertainment) and eudaimonic motivations (that challenge one's way of seeing the world, and focus on meaningful human conditions). Oliver and Bartsch (2010) found a similar split between hedonic and eudaimonic motivations for viewing films, as well as a third factor, suspense, which corresponds to arousal experienced in suspenseful films.

Hartmann (2013) recently reworked these different potential motivations into a framework, which suggests users turn to media for two reasons: recreation or psychological growth. Recreational media use is predominantly characterized by hedonic motivations, whereas psychological growth is characterized by eudaimonic motivations. What is new about Hartmann's conceptualization, however, is that both recreation and psychological growth are conceptualized in terms of the challenge provided to the user by the media, suggesting that recreational media use is less taxing on self-regulatory resources (and indeed, can restore depleted resources, see Reinecke, Klatt, & Krämer, 2011). In contrast, "growth" media require self-regulatory resources by the user to view potentially disturbing or emotionally difficult material in the name of psychological growth and long-term eudaimonic well-being.

While Hartmann (2013) conceptualized the broader nature of challenge in terms of self-regulatory capacity, Bartsch and Hartmann (2015) tested how two types of challenge— cognitive and affective challenge—affect users' entertainment experience (i.e., fun, suspense, appreciation). They found that movies with less challenge of either type (low cognitive and affective challenge) were perceived as fun and, thus, as light-hearted entertainment suitable for recreation, whereas movies high in either cognitive or affective challenge triggered users' appreciation and, thus, were more closely linked to the "growth" or eudaimonic motivations of meaning, lasting worth, and import. Therefore, we turn now to defining what challenge may look like in media, and how it may relate to the self-regulatory capacity of users.

Cognitive and Affective Challenge

Bartsch and Hartmann (2015) discuss several potential sources of cognitive challenge in film content. According to this view, cognitive challenges may build on the difficulty to integrate new information into existing cognitive schemas (Silvia, 2005). For example, the movie *Memento* is filmed in reverse, requiring viewers to keep track of several alternate plot lines and possible futures. The number of plots running concurrently is another example of cognitive load induced by movies or films or television, as Johnson (2006) suggests. He posits that the multiple concurrent plot lines in serial drama such as *The Sopranos* make these series more cognitively challenging than television dramas of the past, which focused on only a few key plot points at a time. This could indicate that complexity is a determinant of cognitive challenge. Cognitive challenges may also result from dissonant information, which is difficult to integrate into an existing schema. For example, a movie may present insights that are not in line with the worldview of most users (e.g., *The Corporation*; *An Inconvenient Truth*) and this dissonance may be perceived as cognitively challenging. Perseverance in the face of these cognitive challenges is likely to draw upon viewers' self-control: Rather than giving in to the desire to expose themselves to more directly rewarding and hedonically pleasant stimuli, cognitively challenging films demand viewers to *endure* thought-provoking and cognitively exhausting content in order to benefit from more complex viewing gratifications, such as appreciation (Oliver & Bartsch, 2010).

Moreover, cognitive challenge may provoke interest in users, especially if it represents a novel or pleasant experience (Silvia, 2005). On the other hand, Silvia

(2005) notes that the interpretation of complex patterns as appealing or interesting depends on the coping resources of the user. In the literature on complexity, cognitive coping resources are linked with perceptual fluency. *Perceptual fluency* refers to the ease with which one can process as well as the efficient processing of a stimulus (Cesario, Corker, & Jelinek, 2013; Winkielman, Schwarz, Fazendeiro, & Reber, 2003). Perceptual fluency is marked by high processing rate, low cognitive resource demands, and high accuracy. As explained by Gillebaart, Förster, and Rotteveel (2012), high perceptual fluency typically signifies a safe environment, which is also hedonically rewarding. In contrast to the warm, safe feeling that familiarity can evoke, novelty or complexity is often associated with threat, as it represents a potential challenge to the organism (Gillebaart et al., 2012). When depleted, users may be less able to cope with potential threats to their self. Therefore, complex or cognitively dissonant media stimuli may be less appealing to depleted users, as they lack the resources to cope successfully with these threats.

In terms of affective challenge, Bartsch and Hartmann (2015) suggest that affective challenges result from the experience of intense negative affect and may be closely related to the sensation of pain or sorrow (e.g., Rozin, 1999). For example, a tragic movie may portray in length the struggle and eventual death of a protagonist that suffers from cancer (e.g., *Beaches*) and may, thus, induce a painfully intense feeling of sadness. Other movies, for example horror movies, may trigger intense levels of fear or disgust in users.

Similar to the processing of cognitive challenges, the processing of affective challenges, too, requires the investment of self-regulatory resources from the user (Baumeister et al., 1998). For example, watching horror films while controlling aversive emotional reactions—showing mastery over the emotional challenge—has been shown to be a method of attracting opposite sex partners (Zillmann, Weaver, Mundorf, & Aust, 1986). Watching a very sad movie provokes self-reflection and the need to overcome or master these touching emotions (Oliver & Bartsch, 2010). This coping response, or even these self-reflective impulses, can draw upon self-regulatory resources. Indeed, a common depletion task is to have viewers watch an emotional video on a screen, and instruct the viewer to *not* mirror any emotions shown in the clip (Gross, 1998). The more intense the challenges imposed by media content, the more effortful it may be for users to process this content in socially or personally acceptable ways.

This said, media that require these types of self-regulatory resources from users should be experienced as more effortful to process. In contrast, the processing of mildly challenging media content should require a low amount of self-regulatory resources. If this is the case, then users with low levels of available self-control (i.e., ego-depleted users) should be drawn to media content requiring fewer resources. Therefore, we pose our first hypotheses in terms of the relative preference for challenge among ego-depleted individuals:

H1: Ego-depleted users are more likely to selectively prefer non-challenging media content versus challenging content.

Ego-depleted users may not just be drawn to less challenging content, however, they may also avoid challenging content. Therefore, our second hypothesis is:

H2: Ego-depleted users are more likely to selectively avoid challenging media content versus non-challenging content.

There is little evidence suggesting what type of media may be most preferred by non-depleted media users. This is because the preference for and attraction to cognitive and affective challenge are likely to be determined by a large number of state and trait variables, among which ego depletion is only one possible explanation. Therefore, while we can make comparative statements in terms of state differences we expect between depleted and non-depleted users, we are less able to make firm statements about the general preferences for challenge among the non-depleted participants. Thus we pose two research questions:

RQ1: Are ego-depleted users less likely to selectively choose challenging media content than non-depleted users?

RQ2: Are ego-depleted users less likely to selectively avoid challenging media content than non-depleted users?

GENERAL METHOD

Overview

The current project tests our hypotheses with two experimental studies: One focusing on cognitively challenging films, and one focusing on affectively challenging films. In both experiments we used the same experimental procedure and analysis plan. The two experiments were run concurrently using the online survey administration software through Qualtrics. All participants were selected from a Dutch university via convenience sampling. Upon clicking the link to the survey, participants were randomly assigned to either Study 1 (cognitive challenge) or Study 2 (affective challenge). First, participants filled out brief demographic information. Next, participants were randomly assigned (within each study) to either the ego-depleted or non-ego-depleted condition. The ego depletion paradigm was developed by Bertrams, Englert, and Dickhäuser (2010). Participants were all instructed to write a short memory about a recent trip they had taken, of at least 10 to 12 sentences, or 700 characters. In both studies, participants randomly assigned to the ego-depleted condition were instructed to leave out the letters *e* and *n*, as these are the most common letters in the Dutch language (Genootschap Onze Taal, 2013). Such overriding of writing habits has previously been demonstrated to deplete self-control strength compared to transcribing a story conventionally (Schmeichel,

2007). Hence, in the non-depletion condition, participants were allowed to use all letters. Afterwards, participants completed two items as a manipulation check asking how difficult the task was. After this assignment the participants had to rank order six movies in terms of which they would prefer to watch immediately, at that moment. These movies were presented in a randomized assortment much like a Netflix choice array and represented three films pre-tested as "high challenge" and three films pre-tested as "low challenge." After rank-ordering their selections, participants were thanked for their time and debriefed.

Stimuli

Cognitive challenge. Twenty-four movies rated as "cognitively challenging" or "cognitively unchallenging" by participants in Bartsch and Hartmann (2015) were presented to 16 respondents to rate as cognitively challenging or unchallenging in a pre-test. Following the procedure used by Bartsch and Hartmann (2015), the movies were rated from 1 (= *no cognitive challenge at all*) to 5 (= *high cognitive challenge*). The movies selected for the high cognitive challenge condition were the highest rated movies of the 24; three science documentaries: *Roving Mars* (M = 3.38, SD = 1.20), *Darwin's Nightmare* (M = 3.69, SD = 1.30), and *What the Bleep Do We Know!?* (M = 3.88, SD = 1.41). For the low cognitive challenge condition the three movies selected were two light-hearted comedies: *The Devil Wears Prada* (M = 2.37, SD = 1.41) and *Bruce Almighty* (M = 2.63, SD = 1.36) and a thriller: *Red* (M = 2.31, SD = 0.95).

Affective challenge. The same 16 respondents also rated 24 films found in Bartsch and Hartmann (2015) to be affectively challenging or unchallenging. Following the procedure used by Bartsch and Hartmann (2015), the movies were rated from 1 (= *no affective challenge at all*) to 5 (= *high affective challenge*). The movies selected for the current study for the high emotional challenge condition were two dramas: *My Sister's Keeper* (M = 4.00, SD = 1.00), *The Pursuit of Happyness* (M = 3.94, SD = 1.20), and one horror movie: *Saw* (M = 3.25, SD = 1.20). For the low emotional challenge condition the movies were two comedies: *American Pie* (M = 2.13, SD = 1.20) and *Playing for Keeps* (M = 2.50, SD = 1.20) and a thriller: *Thinner* (M = 2.44, SD = 1.20).

Measures

Manipulation check. Participants were asked how difficult they found the writing task and how much effort they put in the task. Both questions were 7-point Likert-type scales (1 = *extremely easy* and 7 = *extremely difficult*) and (1 = *no effort* and 7 = *extreme effort*).

Preference for challenge. In both studies, in order to assess selection preference, the movies ranked as first choice were coded as belonging either to the "high challenge" or "low challenge" groups noted above. For example, in the cognitive challenge study, if a participant ranked *Roving Mars, Darwin's Nightmare*, or *What the Bleep Do We Know* as a first choice, they would receive 1 for preference for challenge; if none of those three were selected, they would receive a 0. The same procedure was also used for the "two most preferred" that is, the combined first and second choices.

Avoidance of challenge. The same process was applied to the "least preferred" selections; such that if a participant ranked *Roving Mars, Darwin's Nightmare*, or *What the Bleep Do We Know* as sixth choice, they would receive 1 for avoiding challenge; if none were selected for either sixth choice, they would receive a 0. The same process was also carried out for the "two least preferred," which included both the fifth and sixth choices.

STUDY 1 — COGNITIVE CHALLENGE

Participants

A total of 64 Dutch participants, 19 men (age M = 24.68; SD = 3.71) and 45 women (age M = 24.40; SD = 5.89) completed the survey. The educational level of the participants was high, with 59.4% having some college education. In total, 30 respondents participated in the ego-depleted condition and 34 in the non-ego-depleted condition. There was no relationship between gender and condition, $\chi^2(1)$ =.360, p =.55.

RESULTS

Manipulation check

An independent-samples t-test was conducted to compare if the treatment was successful. For the difficulty item, there was a significant difference in the scores for ego-depleted condition (M = 5.33, SD= 1.71) and non-ego-depleted condition (M = 2.94, SD = 1.58); $t(62)$ = -5.83, p = <.01. Participants also reported having put more effort into the task in the ego-depleted condition (M = 4.27, SD= 1.66) as compared to the non-ego-depleted condition (M = 3.00, SD = 1.13); $t(62)$ = -3.61, p <.01. These results suggest that the ego depletion manipulation was successful.

Analysis

To test our hypotheses, we examined the relative percentages of preference and avoidance of high and low challenge films for both ego-depleted and non-ego-depleted conditions. Although chi-square tests are presented, given that the rankings were comparative versus absolute, (i.e., the movies were only preferred or avoided compared to the other movies presented) the percentages may be more informative than the chi-square tests (see Table 7.1).

Table 7.1. Cognitive Challenge and Ego Depletion.

	Preference: Cognitively challenging movie ranked 1st		Preference: Cognitively challenging movie ranked 1st or 2nd		Avoidance: Cognitively challenging movie ranked 6th		Avoidance: Cognitively challenging movie ranked 6th or 5th	
	No	Yes	No	Yes	No	Yes	No	Yes
Ego-Depleted	.655	.344	.655	.344	.483	.517	.138	.862
Non-Ego-Depleted	.50	.50	.441	.559	.471	.529	.121	.879

Note. This table shows selective exposure to cognitively challenging or non-challenging content as a function of ego depletion. Percentages represent the share of participants within the respective experimental condition who have made the respective media choice.

Within the group of ego-depleted people, a higher percentage chose a low cognitive challenge movie (65.5%) as their top choice than a high cognitive challenge movie (34.5%). In contrast, within the group of more non-depleted people, there was an equal split between high challenge (50.0%) and low challenge movie selection (50.0%). The overall chi-square was not significant, $\chi^2(1) = 1.54, p =.21$.

Examining the top two choices (selection 1 and 2 combined), non-depleted people again had an almost 50/50 split in terms of preference for a high cognitive challenge (44.1%) versus a low cognitive challenge (55.9%) movie combining both first and second choices. Ego-depleted individuals, however, showed more preference for low challenge (65.5%) than high-challenge films (34.5%), $\chi^2(1) = 2.89, p <.10$.

In terms of cognitive challenge avoidance, the least preferred choice among ego-depleted people was slightly more often a high challenge (51.7%) than a low challenge (48.3%) movie. This pattern is also reflected in the choices of non-depleted individuals, where a high cognitive challenge movie was slightly more often the least preferred choice (52.9%). The overall chi-square, however, was not significant, $\chi^2(1)$ =.009, p =.90.

Examining the two least-preferred options, among ego-depleted participants, at least one high cognitive challenge film was included in the two least preferred places (ranked 5th and 6th) for the majority of participants (86.2%). This percentage was almost equal within the group of non-depleted people, in which 87.9% of the two least preferred choices included at least one cognitively challenging movie, $\chi^2(1)$ =.038, p =.85.

STUDY 2—AFFECTIVE CHALLENGE

Participants

A total of 66 Dutch participants completed the experiment. In total, 21 men (M = 28.35; SD = 11.52) and 45 women (M = 23.91; SD = 3.33) filled out the questionnaire. The educational level of the participants was high, with 55% completing some part of a university degree. In total, 32 respondents participated in the ego-depleted condition and 34 in the non-ego-depleted condition. Gender again was unrelated to condition, $\chi^2(1)$ = 1.75, p =.18.

RESULTS

Manipulation check

There was a marginally significant difference in the scores for the ego-depleted condition (M = 4.31, SD = 1.64) and the non-ego-depleted condition (M = 3.47, SD = 1.90); $t(64)$ = -1.927, p <.06. There was a significant difference in the scores for ego-depleted condition (M = 4.84, SD = 1.51) and non-ego-depleted condition (M = 3.91, SD = 1.51) in terms of effort, $t(64)$ = 2.514, p <.05. These results suggest that the manipulation of ego depletion was effective.

Analysis

The results are summarized in Table 7.2. Within the group of ego-depleted people, a higher percentage chose a low affective challenge movie (58.1%) as their top choice than a high affective challenge movie (41.9%). In contrast, within the non-depleted condition, a higher percentage chose a high challenge movie (76.5%) as their top choice than a low challenge movie (23.5%), $\chi^2(1)$ = 8.058, p <.01.

Table 7.2. Affective Challenge and Ego Depletion.

	Preference: Emotionally challenging movie ranked 1st		Preference: Emotionally challenging movie ranked 1st or 2nd		Avoidance: Emotionally challenging movie ranked 6th		Avoidance: Emotionally challenging movie ranked 6th or 5th	
	No	Yes	No	Yes	No	Yes	No	Yes
Ego Depleted	.581	.419	.129	.871	.419	.581	.65	.935
Non-Ego-Depleted	.235	.765	.88	.912	.647	.353	.235	.765

Note. This table shows selective exposure to emotionally challenging or non-challenging content as a function of ego depletion. Percentages represent the share of participants within the respective experimental condition who have made the respective media choice.

However, looking at the top two choices, ego-depleted people almost always included at least one high affective challenge movie in their two top choices (87.1%). This choice pattern was almost identical to that of the non-depleted condition (91.2%), $\chi^2(1) = 0.28, p = .596$.

The least preferred choice among ego-depleted people was slightly more often a high affective challenge (58.1%) than a low challenge (41.9%) movie. In addition, a high challenge movie was more often the least preferred choice within the group of ego-depleted people (58.1%) than within the group of non-depleted people (35.3%), $\chi^2(1) = 3.38, p = .066$.

Among ego-depleted people, the two least preferred choices almost always (93.5%) included at least one affectively challenging movie. This percentage was notably higher than in the group of non-depleted people, in which only 76.5% of the two least preferred choices included at least one affectively challenging movie, $\chi^2(1) = 3.63, p = .057$.

DISCUSSION

In the current studies we tested the notion that ego depletion would affect media choice of cognitively and affectively challenging films. In line with expectations, our results provide preliminary evidence that ego depletion is involved in both selective preference for low challenge films and selective avoidance of high challenge films. This can be seen in the varying percentage of the time ego-depleted persons selected a high challenge versus low challenge film, as well as in comparing ego-depleted to non-ego-depleted participants' selection patterns. Across both cognitive and affective challenge conditions, we see that ego depletion steers people toward less challenging film content and away from challenging film content. Although these patterns were in some cases similar for non-depleted participants, the key finding is that low state self-regulatory capacity may be related to preference for low challenge films and avoidance of high challenge films.

While these results were in line with expectations, this research is still at the exploratory stage. There were limitations to the current study, namely in the selection and ranking of the media stimuli, which in turn limit the conclusions we may draw from our findings, while not detracting from the basic idea. First, in the interest of time, we did not ask for participants' own knowledge of the films used or participants' media habits and preferences. However, this presents confounds when examining choice and preference data, as habit has been shown to be a strong predictor of media choice processes (Ozkaya, 2014). Second, our ranking metric was comparative rather than evaluative. That is, people chose the films they would like to see based on the given array, but not which media they would actually like to see if given unlimited preference and choice. In the interest of brevity, we only offered six films in each study from which to choose. Given the results from Panek (2014), it seems plausible that at the very least including multiple genre or formats would be a better way to offer choice to individuals when examining media selection behaviors than a simple film choice array. This point is related to the third limitation. Due to the ranking procedure, we interpreted our results primarily using the descriptive rather than inferential statistics. The small sample and subsample sizes and the related high uncertainty (i.e., standard error) of the statistics used provide another serious concern. However, the fact that we found a consistent pattern in the descriptive results that were in line with what we expected speaks against the claim that we just relied on "random or inaccurately estimated numbers." Still, from a strict methodological point of view, these results can only be seen as very preliminary (i.e., different samples may indeed provide different percentages) and they must be replicated in the future.

These limitations noted, the results from this research are promising both for understanding the relationship of self-regulation and choice behavior, as well as formulating the notion of challenge in media content as an important determinant of choice. Our findings speak clearly to two areas of theoretical importance for future investigation: Defining challenge in terms of content and selection processes and relating the self-regulatory perspective to media use.

CONCEPTUALIZING MEDIA CHALLENGE

First, it is clear that challenge is an important content factor that has been underutilized in previous media selection research. Although Hartmann (2013) and Bartsch and Hartmann (2015) are beginning to investigate this area, there is more work to be done. As noted in the introduction, challenge may be considered a feature of the content or of the interaction between content and appraisal processes of the viewer. We took a viewer-based approach in this study—using past viewer-based ratings of perceived challenge to define our high and low challenge conditions. Although we posit that challenge relates to self-regulatory resources required in viewing, perceptions of challenge may be driven by various constructs independent of self-regulatory resources. For example, it is clear that content plays a role in the perception of challenge; the cognitively challenging films were all documentaries, and the emotionally challenging films were drama or thriller types. In the current research, we did not take into account these structural features that may also contribute to a perception of challenge in content, such as the genre of the film, the number of concurrent plot threads, the number of actors or main characters to follow, or the storytelling devices used (i.e., narrative complexity), or indeed how these may relate to overall challenge. Thus, a careful examination of what types of structural features promote perceived challenge in viewers is of key interest in future work.

Considering the media in terms of self-regulatory processes, it may be also worth considering the social environment in terms of how challenging media can be. Maybe it is not so much the actual challenge of media content but its reputation as being challenging or recreational that drives the perception of challenge. Maybe some guilty-pleasure media (e.g., violent or sexually explicit content) are "forbidden fruits" (because they are socially undesirable) that are particularly hard to resist for ego-depleted viewers. Similarly, media with content that requires a certain lack of social inhibition to enjoy (such as racially-charged or stereotype-driven humor, for example) may actually be more enjoyable to depleted individuals. If, as Freud suggests, we enjoy humor for its ability to relieve us of the pressure of conforming to social norms, then the appreciation of socially inappropriate humor may be heightened when viewers lack self-regulatory resources (e.g., Muraven, 2008).

Finally, the media form and situation of use may be of interest in conceptualizing challenge. As Panek (2014) noted, only the use of online videos and social network sites was related to low self-control in users, whereas television and DVD use was not. Panek (2014) theorized that the ubiquitous nature of online videos and social networking sites—available via phone, laptop, or tablet—may have contributed to the heightened desirability of these services to users low in self-control. As Reinecke et al. (2014) discuss, media are characterized by having a low threshold for participation—we can watch online videos while also completing other tasks, for example. It may be that some media are challenging not because of content or perceived social norms, but instead because they simply take more time to watch. Committing to a multi-season program can take hundreds of hours of viewing time, whereas watching a YouTube video takes, at most, hundreds of minutes. Therefore, the form of media may also require self-regulatory resources.

Taking these different perspectives into account, future research must conceptualize challenge both from a content- and a user-based perspective and isolate the causes of perceived challenge in a way that generalizes across media content, viewing forms, and conditions.

Self-Regulatory Selection vs. Mood-Management Theory

Moving from content to selection processes, a clear definition of challenge in terms of self-regulation will be most beneficial in determining how self-regulatory challenge may fit into previous conceptualizations of media selection. Most past selection work has focused on the hedonic management potential of media, such as mood management theory (MMT; Zillmann, 1988). MMT assumes that individuals strive for the minimization of exposure to negative stimuli while also aiming to maximize exposure to pleasurable stimuli. The theory proposes that individuals arrange stimuli in their environment in a way that optimizes their chances of reaching these goals, and the selection of entertaining media is one form of such arrangement. The theory puts emphasis on four characteristics of individuals and of media stimuli that are crucial for mood management: *excitatory homeostasis*, the selection of media to achieve optimal levels of arousal; *intervention potential*, the ability of a message to intervene in some individual's attention; *hedonic valence*, the emotional tone of media stimuli; and *behavioral affinity*, or the similarity of media content to real-life situations resulting in the current mood state (Bryant & Davies, 2006).

Reinecke et al. (2012) proposed recently that the four mechanisms of mood management might be thought of as representing two processes: 1) those that simply distract an individual from negative mood versus 2) those that address its cause through repair. Conceptually, hedonic valence, intervention potential, and message-behavioral affinity could be understood in terms of distraction processes. Alternatively, excitatory homeostasis can be understood in terms of repair processes. That is, negative mood caused by low levels of arousal can be addressed by exposing oneself to highly arousing media (repair), whereas negative mood caused by, e.g., frustration, can be addressed by exposing oneself to highly absorbing content (distraction).

There are clear difficulties reconciling a self-regulatory perspective and a mood-management perspective in terms of media choice. The leading question posed by MMT is: What media do people choose if they are in a negative mood, overly aroused or bored? That is, MMT explains media choice based on people's mood and excitatory states. A self-regulation approach, in contrast, asks: What media do people choose if they are tired/exhausted? That is, the second perspective approaches media choice based on a lack of self-regulatory resources or ego depletion. Ego depletion is not the same as boredom, and also not negative mood although we know that ego depletion may be slightly related to negative affect (Hagger, Wood, Stiff, & Chatzisarantis, 2010). Clearly, self-regulatory demand or "potential amount of self-regulatory energy needed to process content and cope with its effects" (as a media variable) is not the same as valence (positive-negative) or arousal potential (arousing-soothing), the two big media variables that are crucial in mood management theory.

This said, there are some noteworthy overlaps between both approaches. For example, ego-depleted people are more prone to seek something "nice," which could indeed sometimes mean pleasurable or exciting (Hagger et al., 2010; Shiv & Fedorikhin, 1999). Pleasure counteracts ego depletion, having a restorative effect (Reinecke et al., 2011), and people are drawn to pleasurable experiences when ego-depleted, perhaps for the hedonic regulation posited by MMT. However, in terms of self-regulatory capacity, pleasure may mean many different things—emotionally stimulating, hedonically pleasant, and so on. Indeed, the affective and arousal state of ego depletion is unknown. Ego depletion may be caused by a variety of inductions such as choice situations (e.g., "choose from among 5, 10, 50 options"), stressful tasks (e.g., work or school participation), self-regulating of emotion or thoughts (e.g., "do not think of a camel"), all of which may produce disparate effects on affect and arousal (see Hagger et al., 2010, for an overview). Therefore, future research may need to first account for the affective and arousal states of ego depletion, before anticipating how this depletion may be best addressed by media use.

Closely connected to the uncertainty concerning the conceptual overlap of selective exposure arising from mood management versus self-regulation is the question whether ego-depleted media users actually tend to make the right choices in terms of media content that facilitates recovery from their depleted self-control capacity. The results from Reinecke et al. (2014) suggest that ego-depleted individuals, who could benefit particularly strongly from media-induced recovery effects, ironically are less likely to experience recovery and media enjoyment due to their negative appraisal of their own media use. Taking this into account, the results of the present research imply that the selective exposure patterns of ego-depleted individuals might be a further factor turning media use into a guilty pleasure. While other leisure time activities, such as sports, social, or cultural activities, are often seen as socially desirable, entertaining media use is often conceptualized as a less desirable or productive activity (e.g., Williams, 2003).

The results of this research demonstrate that ego-depleted individuals are particularly drawn to emotionally and cognitively non-challenging media stimuli. Such "lowbrow" media content is likely to be perceived as less socially desirable than more thought-provoking forms of "meaningful entertainment" (Oliver & Raney, 2011, p. 985). Ego-depleted media users may thus not only feel guilty for their media use because it conflicts with other goals, but also because they tend to select forms of media content that are particularly socially undesirable. These negative side effects of their selection patterns, in turn, reduce their chance to benefit from their media use in terms of recovery and enjoyment (Reinecke et al., 2014). In conclusion, while the occasional couch-potato evening is surely no public health concern, a nation of fatigued, pleasure-seeking individuals shying away from deeper engagement with meaningful narrative in favor of hedonic "brain candy" surely is. As media entertainment continues to be more available and accessible, understanding how and when individuals turn to what kind of content becomes even more important. Further disentangling the complex relationship between state self-control, selective exposure processes, and the resulting effects on psychological well-being, thus remains a highly relevant challenge for future research and practice.

REFERENCES

Bartsch, A., & Hartmann, T. (2015). The role of cognitive and affective challenge in entertainment experience. *Communication Research*. Advance online publication. doi: 10.1177/0093650214565921

Baumeister, R. F., & Heatherton, T. F. (1996). Self-regulation failure: An overview. *Psychological Inquiry, 7*(1), 1–15.

Baumeister, R. F., Bratslavsky, E., Muraven, M., & Tice, D. M. (1998). Ego depletion: Is the active self a limited resource? *Journal of Personality and Social Psychology, 74*(5), 1252–1265. doi:10.1037/0022–3514.74.5.1252

Baumeister, R. F., Vohs, K. D., & Tice, D. M. (2007). The strength model of self-control. *Current Directions in Psychological Science, 16*(6), 351–355. doi:10.1111/j.1467–8721.2007.00534.x

Bertrams, A., Englert, C., & Dickhäuser, O. (2010). Self-control strength in the relation between trait test anxiety and state anxiety. *Journal of Research in Personality, 44*(6), 738–741.

Brosius, H. B., Rossmann, R., & Elnain, A. (1999). Alltagsbelastung und Fernsehnutzung. Wie beeinflußt der Tagesablauf von Rezipienten den Umgang mit Fernsehen. *Hasebrink, Uwe/Rössler, Patrick (Hg.): Medienrezeption zwischen Individualisierung und Integration. München: R. Fischer*, 167–186.

Bryant, J., & Davies, J. (2006). Selective exposure processes. In J. Bryant and P. Vorderer (Eds.), *Psychology of Entertainment*, (pp. 19–33). Mahwah, NJ: Erlbaum.

Cesario, J., Corker, K. S., & Jelinek, S. (2013). A self-regulatory framework for message framing. *Journal of Experimental Social Psychology, 49*(2), 238–249.

Gailliot, M. T., Baumeister, R. F., DeWall, C. N., Maner, J. K., Plant, E. A., Tice, D. M., & Schmeichel, B. J. (2007). Self-control relies on glucose as a limited energy source: Willpower is more than a metaphor. *Journal of Personality and Social Psychology, 92*(2), 325–336.

Genootschap Onze Taal. (2013). *Letterfrequeuntie in het Nederlands*. Retrieved from http://onzetaal.nl/taaladvies/advies/letterfrequentie-in-het-nederlands

Gillebaart, M., Förster, J., & Rotteveel, M. (2012). Mere exposure revisited: The influence of growth versus security cues on evaluations of novel and familiar stimuli. *Journal of Experimental Psychology: General, 141*(4), 699–714.

Gross, J. J. (1998). Antecedent-and response-focused emotion regulation: divergent consequences for experience, expression, and physiology. *Journal of Personality and Social Psychology, 74*(1), 224.

Hagger, M. S., Wood, C., Stiff, C., & Chatzisarantis, N. L. (2010). Ego depletion and the strength model of self-control: A meta-analysis. *Psychological Bulletin, 136*(4), 495–525.

Hartmann, T. (2013). Media entertainment as a result of recreation and psychological growth. *The International Encyclopedia of Media Studies, 5*, 1–7.

Hofmann, W., Vohs, K. D., & Baumeister, R. F. (2012). What people desire, feel conflicted about, and try to resist in everyday life. *Psychological Science, 23*, 582–588. doi:10.1177/0956797612437426

Johnson, S. (2006). *Everything bad is good for you.* New York, NY: Riverhead.

Muraven, M. (2008). Prejudice as self-control failure. *Journal of Applied Social Psychology, 38*(2), 314–333.

Muraven, M., & Baumeister, R. F. (2000). Self-regulation and depletion of limited resources: Does self-control resemble a muscle? *Psychological Bulletin, 126*(2), 247–259.

Oliver, M. B., & Bartsch, A. (2010). Appreciation as audience response: Exploring entertainment gratifications beyond hedonism. *Human Communication Research, 36*(1), 53–81.

Oliver, M. B., & Raney, A. A. (2011). Entertainment as pleasurable and meaningful: Differentiating hedonic and eudaimonic motivations for entertainment consumption. *Journal of Communication 61*, 984–1004. doi: 10.1111/j.1460–2466.2011.01585.x

Ozkaya, E., (2014, May). *The role of habit and emotional regulation on entertainment video selection.* Paper presented at the Sixty-fourth International Communication Association Conference, Seattle, WA.

Panek, E. (2014). Left to their own devices: College students' "guilty pleasure" media use and time management. *Communication Research, 41*(4), 561–577.

Reinecke, L. (2009). Games and recovery: The use of video and computer games to recuperate from stress and strain. *Journal of Media Psychology: Theories, Methods, and Applications, 21*(3), 126–142.

Reinecke, L., Hartmann, T., & Eden, A. (2014). The guilty couch potato: The role of negative emotions in reducing recovery through media use. *Journal of Communication, 64*(4), 569–589. doi: 0.1111/jcom.12107

Reinecke, L., Klatt, J., & Krämer, N. C. (2011). Entertaining media use and the satisfaction of recovery needs: Recovery outcomes associated with the use of interactive and noninteractive entertaining media. *Media Psychology, 14*, 192–215. doi: 10.1080/15213269.2011.573466

Reinecke, L., Tamborini, R., Grizzard, M., Lewis, R., Eden, A., & Bowman, N. D. (2012). Characterizing mood management as need satisfaction: The effects of intrinsic needs on selective exposure and mood repair. *Journal of Communication, 62*(3), 437–453.

Rozin, P. (1999). Preadaptation and the puzzles and properties of pleasure. In D. Kahneman, E. Diener, & N. Schwarz (Eds.), *Well-being: The foundations of hedonic psychology* (pp. 109–133). New York, NY: Russell Sage.

Schmeichel, B. J. (2007). Attention control, memory updating, and emotion regulation temporarily reduce the capacity for executive control. *Journal of Experimental Psychology: General. 136*(2), 241–255.

Shiv, B., & Fedorikhin, A. (1999). Heart and mind in conflict: The interplay of affect and cognition in consumer decision making. *Journal of Consumer Research, 26*(3), 278–292.

Silvia, P. J. (2005). Cognitive appraisals and interest in visual art: Exploring an appraisal theory of aesthetic emotions. *Empirical Studies of the Arts, 23*(2), 119–133.

Vorderer, P., & Kohring, M. (2013). Permanently online: A challenge for media and communication research. *International Journal of Communication, 7*, 188–196.

Vorderer, P., Klimmt, C., & Ritterfeld, U. (2004). Enjoyment: At the heart of media entertainment. *Communication Theory, 14*(4), 388–408.

Wagner, D. T., Barnes, C. M., Lim, V. K., & Ferris, D. L. (2012). Lost sleep and cyberloafing: Evidence from the laboratory and a daylight saving time quasi-experiment. *Journal of Applied Psychology, 97*(5), 1068–1076.

Williams, D. (2003). The video game lightning rod: Construction of a new media technology, 1970–2000. *Information, Communication & Society, 6,* 523–550. doi: 10.1080/1369118032000163240

Winkielman, P., Schwarz, N., Fazendeiro, T., & Reber, R. (2003). The hedonic marking of processing fluency: Implications for evaluative judgment. In J. Musch & K. C. Klauer (Eds.), *The psychology of evaluation: Affective processes in cognition and emotion* (pp. 189–217). Mahwah, NJ: Erlbaum.

Zillmann, D. (1988). Mood management through communication choices. *American Behavioral Scientist, 31,* 327–340.

Zillmann, D., & Bryant, J. (1985). Affect, mood, and emotion as determinants of selective exposure. In D. Zillmann & J. Bryant (Eds.), *Selective Exposure to Communication* (pp. 157–189). Hillsdale, NJ: Erlbaum.

Zillmann, D., Weaver, J. B., Mundorf, N., & Aust, C. F. (1986). Effects of an opposite-gender companion's affect to horror on distress, delight, and attraction. *Journal of Personality and Social Psychology, 51*(3), 586–594.

CHAPTER EIGHT

The Secret to Happiness

Social Capital, Trait Self-Esteem, and Subjective Well-Being

JIAN RAYMOND RUI
TEXAS TECH UNIVERSITY, USA

MICHAEL A. STEFANONE
UNIVERSITY AT BUFFALO, STATE UNIVERSITY OF NEW YORK, USA

The history of humanity has shown that we work hard to pursue a good life and individual well-being. For example, personal wealth is closely related to well-being (Dowrick, 2004) and many technologies are developed to generate more wealth. Over the past ten years, web-based technologies such as social media gained increasing popularity. Most of these technologies are designed to help individuals improve their interpersonal relationships, which is an important component of well-being (Alkire, 2002).

It is still unclear, however, what impact technology is having on interpersonal relationships. If web-based communication technologies are improving the quality of our social networks and interpersonal communication, you would hypothesize that broad-level indicators of well-being should also be improving. Consider the suicide rate in North America as an example. This rate should be negatively related to well-being (McGillivray, 2007). However, according to the Centers for Disease Control and Prevention, 38,364 suicides were reported in the U.S. in 2010, equivalent to12.1 deaths per 100,000 people (American Foundation for Suicide Prevention, n.d.), which suggest almost no change compared with the rate of 12.3 deaths per 100,000 people in 1981.

This chapter reports on a study aimed at better understanding individual well-being by explicating a set of specific factors that affect well-being. One approach to defining well-being is based on individuals' evaluation of subjective happiness (Kahneman, Diener, & Schwarz, 1999; Ryan & Deci, 2001). Researchers argued that subjective happiness was the key component to defining well-being and proposed subjective well-being (Diener, Sapyta, & Suh, 1998).

Scholarship suggests a state-trait framework to explain human behavior (Steyer, Schmitt, & Eid, 1999), which also applies to subjective well-being. While state variables are transient, trait variables are relatively stable. Research shows that these two types of variables interact to predict human behavior.

One important state variable that consistently exhibits a positive relationship with subjective well-being is social capital (Helliwell, 2007; Ko & Kuo, 2009; Putnam, 2000). However, the process by which social capital affects subjective well-being is unclear in the extant literature. Personal social ties provide a potential approach to explaining this question as social capital is embedded in human relations. Research shows that there are two types of social capital: bonding and bridging (Putnam, 2000), and they are originated from strong and weak ties, respectively (Wellman & Wortley, 1990; Williams, 2006). As a result of this difference, bonding and bridging social capital may have different relationships with subjective well-being. Empirical evidence to support this argument is lacking. However, understanding how bonding and bridging social capital are related to subjective well-being can contribute to the scholarship about subjective well-being from the perspective of personal networks.

Furthermore, trait variables such as psychological, emotional, and cognitive conditions are also related to subjective well-being (Argyle & Lu, 1990; DeNeve & Cooper, 1998; Diener, Larson, & Emmons, 1984). Therefore, social capital may be intertwined with trait variables to affect subjective well-being. Specifically, we propose that social capital and trait self-esteem may work together to influence subjective well-being because trait self-esteem affects how one makes attribution of outside events (Fitch, 1970).

The literature review of this study is structured as follows. We start with defining subjective well-being. Next, we discuss the definition of social capital, how bonding and bridging social capital are different and related to subjective well-being. Finally, we explain how trait self-esteem is intertwined with bonding and bridging social capital to influence subjective well-being.

SUBJECTIVE WELL-BEING, SOCIAL CAPITAL, AND SELF-ESTEEM

Subjective Well-Being

The Cyrenaic school of philosophy in ancient Greece argues that the ultimate goal of life is to seek and maximize happiness, which can be obtained from physical pleasure (Ryan & Deci, 2001). While this narrows the meaning of well-being to physical enjoyment, more recent scholarship extends it to broader accomplishments.

Diener et al. (1998) argued that happiness could be derived from accomplishing personal goals or achieving valued outcomes in desired realms. Although these two perspectives differ in what happiness involves, they both agree that subjective evaluation of one's feeling is key to defining well-being. This constitutes the basis of conceptualizing subjective well-being. In the present study, following prior work (e.g., Diener & Lucas, 1999), we focus on subjective well-being and define it as subjective happiness.

As individuals' subjective feeling is key to subjective well-being, operationalization focuses on individual evaluation of their affect and life (McGillivray, 2007). Empirical measures generally involve two components: life satisfaction (Diener, Emmons, Larsen, & Griffin, 1985; Pavot & Diener, 1993), and the balance between positive and negative affect (Diener et al., 2010; Watson, Clark, & Tellegen, 1988). However, Schwarz and Strack (1999) argued that the global measure of life satisfaction could be influenced by individual mood at the moment of responding to the scale. Thus, this measure is subject to the risk of reducing global life satisfaction to state-level affect. In addition, the conceptual definition of subjective well-being highlights the importance of affect balance to defining subjective well-being. Thus, in the present study, we operationalize subjective well-being as the balance between negative and positive affect, following Diener et al. (2010).

Social Capital and Subjective Well-Being

The state-trait theoretical framework suggests that stable, internal variables and transient, external variables combine to influence human behavior (Steyer et al., 1999). In terms of subjective well-being, prior research shows that a wide range of state- and trait-level variables can contribute to subjective well-being. Examples of state variables involve individuals' social network (Gundelach & Kreiner, 2004; Yip et al., 2007), personal relationships with others (Helliwell & Putnam, 2004; Myers, 2000), and income (Haring, Stock, & Okun, 1984). Examples of trait variables include personality (Costa & McCrae, 1980), self-esteem (Dunning, Leuenberger, & Sherman, 1995; Kwan, Bond, & Singelis, 1997), and perceptions of one's health condition (Diener, Suh, Lucas, & Smith, 1999). One important state variable that generally exhibits a positive relationship with subjective well-being in prior research is social capital (Helliwell, 2007; Ko & Kuo, 2009; Putnam, 2000).

Prior research generates different definitions of social capital. For instance, Onyx and Bullen (2000) defined social capital as the process by which strong relationships characterized by trust and reciprocal help are developed, whereas other research highlights the structural features of social network (e.g., Lin, 1982) and membership in social organizations (e.g., Putnam, 2000) in social capital. In the present study, we follow Bourdieu and Wacquant (1992)'s definition and conceptualize it as the resources that individuals can gain from their social contacts because the value of social capital lies in its ability to generate resources that can help individuals (Resnick, 2002).

Putnam (2000) distinguishes two types of social capital: bonding and bridging, and they are associated with different personal ties. Bonding social capital is originated from strong ties such as family and close friends, whereas bridging social capital is generated from weakly connected individuals like acquaintances (Williams, 2006). Strong ties involve emotional attachment and relational closeness, so they are more willing to devote time and effort to helping each other. As a result, bonding social capital includes emotional support and financial support, which require time and effort (Wellman & Wortley, 1990). However, one drawback of bonding social capital is that strong ties are usually composed of similar individuals (Granovetter, 1973) and thus can generate redundant information that the ego already knows.

On the other hand, bridging social capital compensates for this weakness. Compared to strong ties, weak ties are more likely composed of social contacts from different backgrounds. Thus, they likely generate resources that individuals usually do not have access to (Granovetter, 1973). Empirical evidence shows that acquaintances are better at providing novel information that generates additional opportunities for career development and promotion compared to family (Granovetter, 1983). Therefore, the strength of bridging social capital lies in its access to novel information (Williams, 2006).

Although social capital is found to have a generally positive relationship with subjective well-being (Helliwell, 2007; Ko & Kuo, 2009; Putnam, 2000), the process by which social capital affects subjective well-being still remains unclear. Specifically, consider that bonding and bridging social capital are originated from different social ties and involve different types of support. Hence, they may function differently to affect subjective well-being.

Recall that subjective well-being refers to subjective evaluation of happiness and is operationalized as the balance between positive and negative affect. Because bonding social capital results from strong ties, individuals are closer and more attached to these contacts. In addition, bonding social capital provides resources that require more time and effort, which weak ties cannot or do not want to afford. Thus, receiving these resources can strengthen the relationship and possibly heighten subjective well-being. On the contrary, emotional closeness and relational strength is lacking between weak ties. The novel informational resources that weak ties can provide involve less investment of time and effort. Therefore, receiving bridging social capital may not increase individuals' happiness. Thus, bonding social capital should be positively related to subjective well-being but bridging social capital may not.

H1: Bonding social capital is positively related to subjective well-being.

Recall that trait- and state-level variables combine to influence subjective well-being. In the present study, we propose that trait self-esteem provides a potential approach to explicating how these two types of factors interact to influence subjective well-being.

Trait Self-Esteem, Social Capital, and Subjective Well-Being

Trait self-esteem refers to individuals' global appraisal of self-worth (Rosenberg, 1989). Research has found trait self-esteem is positively associated with positive affect, life satisfaction, and perceptions of social relationships (Fox, 1997; Furnham & Cheng, 2000; Lyubomirsky, Tkach, & Dimatteo, 2006). These findings suggest that trait self-esteem can be a valuable approach to explicating the process by which social capital affects subjective well-being.

Recall that bonding social capital is associated with resources like emotional and financial support, which require investment of time and effort in those relationships. Therefore, when individuals provide these resources, it signals recognition that the recipients of these resources are important and the relationships with recipients are valued. Thus, by receiving bonding social capital, individuals subsequently feel loved and respected. Over time, these positive feelings can in turn increase trait self-esteem (Bishop & Inderbitzen, 1995; Ryan, Stiller, & Lynch, 1994; Symister & Friend, 2003). As trait self-esteem is positively related to positive affect (Valkenburg, Peter, & Schouten, 2006), the relationship between bonding social capital and subjective well-being should be mediated by trait self-esteem.

H2: Trait self-esteem mediates the relationship between bonding social capital and subjective well-being.

In addition, although we propose no direct relationship between bridging social capital and subjective well-being, trait self-esteem may be a moderator between them. This is because individuals with different levels of trait self-esteem make different causal attributions to outside events. While high trait self-esteem individuals attribute negative events to external factors as a coping strategy, low trait self-esteem individuals tend to make internal attributions (Fitch, 1970; Tennen & Herzberger, 1987). In the present study, receiving bridging social capital from weak ties is inherently positive, but the relatively low investment associated with providing bridging social capital can make low trait self-esteem individuals feel less recognized or loved. In contrast, high trait self-esteem individuals may consider receiving bridging social capital as recognition of self-worth and evidence that they have positive relationships with support providers. Therefore,

H3: Trait self-esteem moderates the relationship between bridging social capital and subjective well-being in that a positive relationship between bridging social capital and subjective well-being is only found among high trait self-esteem individuals.

METHOD

Participants

As part of a larger project, an online survey was conducted at a large Northeastern university in spring 2013. The sample was recruited from college students taking introductory communication classes. An announcement of the survey was made both in class and on the course website, where students could find the link to the survey. All research procedures were approved by the institutional review board. A total of 223 responses were collected, 49.5% were male; 23.9% were freshmen, 38.7% were sophomores, 29.7% were juniors, and 7.7% were seniors; 63.5% were Caucasian, 19.8% were Asian, 10.4% were African American, and 4.0% were Hispanic.

Measures

Bonding and *bridging* social capital were measured with Ellison, Steinfield, and Lampe's (2011) Likert scale, which includes five (M = 3.14, SD = .90, Cronbach's α = .76) and six items (M = 3.32, SD =.92, Cronbach's α =.90), respectively.

Trait self-esteem was measured with Rosenberg's 7-item Likert scale (1989) that assesses participants' general evaluations of self (M = 3.93, SD = .83, Cronbach's α = .89).

Subjective well-being was measured with Diener et al. (2010)'s scale. Participants were asked to report the extent to which they recently experienced six positive emotions and six negative emotions on a 5-point Likert scale (1 = *very rarely or never*, 2 = *rarely*, 3 = *sometimes*, 4 = *often*, 5 = *very often or always*), which was then indicated by subtracting scores on negative emotions from positive ones (M = 6.73, SD = 7.75, Cronbach's α =.68).

RESULTS

Table 8.1. Descriptive Statistics and Zero-Order Correlations for Variables; Means (Standard Deviations) Presented Along the Diagonal.

	Bonding social capital	Bridging social capital	Trait self-esteem	Subjective well-being
Bonding social capital	3.14 (.90)	.63**	.33**	.33**
Bridging social capital		3.32 (.92)	.38**	.32**
Trait self-esteem			3.93 (.83)	.67**
Subjective well-being				6.73 (7.75)

Note: ** p < .01.

Table 8.1 shows the descriptive statistics of and bivariate correlations between variables in this study. In order to test H1 and H2, we employed Hayes' (2013) process analysis. Our model includes subjective well-being as the dependent variable, bonding social capital as the independent variable, trait self-esteem as the mediator, controlling for sex and year in school. This model was estimated for 10,000 bootstrapped samples.

Results first showed a positive relationship between bonding social capital and trait self-esteem ($B = .31$, $p < .001$), $R^2 = .11$, $F(3, 218) = 9.29$, $p < .001$). Second, before trait self-esteem was entered, bonding social capital was positively related to subjective well-being ($B = 2.78$, $R^2 = .12$, $F(3, 218) = 9.9$, $p < .001$). This relationship remained significant but became weak after trait self-esteem was entered ($B = .95$, $p < .05$; adjusted $R^2 = .47$, $F(4, 217) = 48.60$, $p < .001$). In addition, the indirect effect between bonding social capital and subjective well-being through trait self-esteem was estimated to lie between 1.07 and 2.70 with 95% confidence, indicating a significant mediation relationship between bonding social capital and subjective well-being through trait self-esteem. Comparing the indirect effect (1.82) with the total effect (2.78) shows that 65.47% of the relationship between bonding social capital and subjective well-being was attributed to trait self-esteem. Therefore, H1 and H2 were supported (see Figure 8.1).

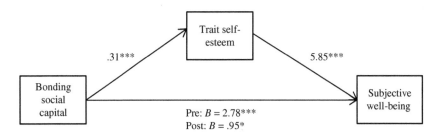

Figure 8.1. Trait Self-Esteem Mediating the Relationship between Bonding Social Capital and Subjective Well-Being.

Next, we conducted ordinary least squares (OLS) regression to test H3. Our model included sex and year in school as control variables, bridging social capital, trait self-esteem, and the interaction between bridging social capital and trait self-esteem as independent variables, and subjective well-being as the dependent variable. Results demonstrated that our model was significant, adjusted $R^2 = .47$, $F(5, 216) = 40.17$, $p < .001$ (see Table 8.2). Trait self-esteem ($\beta = .06$, $p < .001$) and the interaction effect between trait self-esteem and bridging social capital ($\beta = .13$, $p < .02$) demonstrated significant relationships with subjective well-being. However, the main effect of bridging social capital on subjective well-being was not found ($p > .05$). In addition, following Aiken and West (1991), we examined the simple effect of bridging social capital on subjective well-being among participants who reported a high level of trait self-esteem (1 standard deviation above the mean) and who reported a low level of trait self-esteem (1 standard deviation below the mean). A significant relationship between bridging social capital and subjective well-being was only found among high trait self-esteem individuals (β = .18, p < .01) but not among low trait self-esteem individuals. Therefore, H3 was supported (see Figure 8.2).

Table 8.2. OLS Model Regressing Subjective Well-Being on Bridging Social Capital and Trait Self-Esteem.

	Subjective well-being	
	B	SE
Gender	-.56	.76
Year in school	-.81	.43
Bridging social capital	.54	.45
Trait self-esteem	6.23***	.51
Interaction, bridging social capital*trait self-esteem	1.19*	.48
F, Adj. R^2	F (5, 216) = 40.17, .47***	

Note: ***p < .001, *p < .05; Male = 0, Female = 1.

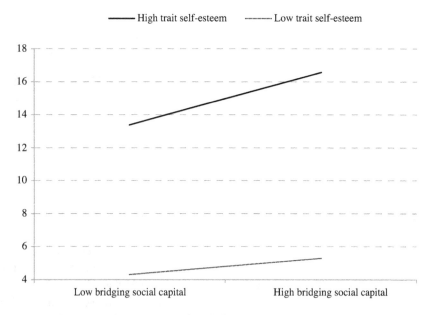

Figure 8.2. Moderation Effect of Trait Self-Esteem (SE) on the Relationship between Bridging Social Capital (BRSC) and Subjective Well-Being (SWB).

DISCUSSION

This study was motivated by the complex and often unclear relationship between technology use, interpersonal relationships, and outcomes like subjective well-being. Drawing upon the classic state-trait framework, we explicated the process by which social capital contributes to subjective well-being, conceptualized as subjective evaluation of happiness and operationalized as the balance between positive and negative affect. Specifically, the present study involves two goals. First, we aim to investigate whether bonding and bridging social capital exhibit different relationships with subjective well-being. Second, we explained how bonding and bridging social capital were intertwined with trait self-esteem to influence subjective well-being. Our findings reveal that only bonding social capital was positively related to subjective well-being. In addition, while trait self-esteem mediated the relationship between bonding social capital and subjective well-being, it moderated the relationship between bridging social capital and subjective well-being. These results demonstrate how individuals' social relationships and their reappraisal of self-worth affect each other to influence their subjective well-being. Together, these results provide theoretical and practical implications to the pursuit of happiness, an everlasting theme of human history.

Major Findings

We first found that bonding and bridging social capital exhibited different relationships with subjective well-being. While bonding social capital was positively related to subjective well-being, bridging social capital was not. This suggests that specific resources are associated with specific ties, consistent with much extant research. Importantly, the provision of different kinds of resources in relationships functionally signals the value of those relationships. For example, an investment like the provision of emotional support suggests that providing bonding social capital also involves more cost, opposed to resources associated with bridging capital. Hence, providing bonding social capital signals the providers' concern for recipients' well-being, and thus makes recipients feel positive about themselves and their relationships. In contrast, such investment is not linked with bridging social capital. Thus, the provision of bridging resources is not associated with increases in individual subjective well-being.

This finding suggests that subjective well-being depends on who to communicate with and what is exchanged between individuals. First, based on our finding, strong ties are more important to subjective well-being. Consider that nowadays new technology expands individuals' personal network, and research shows that most Internet-based technologies are especially effective to expanding weak-tie networks (Ellison et al., 2011). However, network size may not help improve subjective well-being; what matters is the quality of social relationships. Strong ties display emotional attachment and relational closeness through the provision of support that requires large investment of time and effort. Only being connected with these ties can make us happy. This does not mean weak ties are meaningless. We acknowledge that bridging social capital generates novel informational resources that can help individuals with career development. However, they do not necessarily help to improve subjective well-being.

In addition to who to communicate with, what is exchanged between individuals also plays an important role in predicting subjective well-being. The reason why bonding social capital can improve one's subjective well-being is that it generates resources that require large investment of time, material, and effort. The magnitude of investment indicates how important support providers perceive recipients and thus can influence relationships between providers and recipients. This suggests that the cost of support provision is a helpful approach to understanding human relations and personal well-being.

Besides finding the direct relationship between social capital and subjective well-being, we also explicated how bonding/bridging social capital and trait self-esteem combined to predict subjective well-being. We found that trait self-esteem mediated the relationship between bonding social capital and subjective well-being. This result reveals the importance of trait self-esteem to subjective well-being. Providing bonding social capital that involves large investment of time, material, and effort indicates support providers' concern over recipients, as stated above. This thus heightens support recipients' self-esteem, which is translated into an increase in subjective well-being. Recall that trait self-esteem accounted for nearly two thirds of the relationship between bonding social capital and subjective well-being. Therefore, although receiving bonding social capital is effective to improving subjective well-being, the key to happiness is trait self-esteem as positive views of self increase one's subjective evaluations of affect.

Furthermore, although the main effect of bridging social capital on subjective well-being was not found, we found that trait self-esteem moderated their relationship. Specifically, the positive relationship between bridging social capital and subjective well-being was only found among individuals with high levels of trait self-esteem. Different attributions associated with different levels of trait self-esteem accounted for this result. High trait self-esteem individuals generally have a more positive outlook on life, so they may perceive receiving bridging social capital more positively and consider it as evidence that they are liked and respected. On the contrary, because bridging social capital involves less investment in time, material, and effort, low trait self-esteem individuals may not value receiving this resource. Therefore, internal individual traits can change perceptions of external factors. Although social relationship is important to happiness, how individuals perceive these relationships is even more important. This result, along with the finding about the mediation role of trait self-esteem in bonding social capital and subjective well-being, suggests the key role of individual perceptions of self and relationships in pursuing happiness.

Limitations

First, our study employed cross-sectional design, so we cannot make causal arguments. This is a limitation of much of the extant research addressing the relationship between perceptions of social capital and outcomes. However, Stefanone, Kwon, and Lackaff (2012) provide evidence that questions the actual relationship between perceptions of social capital and actual, behavioral outcomes. Their results did not demonstrate a relationship between perceptions of bridging and bonding capital, and resource provision (*enacted* social support).

Second, we used college students as our sample, which limits the generalizability of our results to the general population. Considering that college students are a generally homogeneous group and homogeneity is related to bridging social capital, college students may exhibit systematically lower levels of bridging social capital.

Practical Implications

Despite these limitations, this study generates following practical implications for counseling services, school education, and campaigns aimed at improving personal well-being. As mentioned above, who to communicate with and what is exchanged between individuals are important to improving subjective well-being. Thus, the first strategy is to develop and maintain strong tie networks. In addition, providing resources that require significant investment such as emotional and financial support is an important technique to maintain strong tie networks and improve subjective well-being. Furthermore, our results also highlight the importance of trait self-esteem to subjective well-being. Thus, encouraging individuals to keep positive views of self-worth is a central task in campaigns promoting personal well-being, school education, and counseling services.

Theoretical Implications

Although it seems self-explanatory that social capital exhibits a positive relationship with subjective well-being, we went a step further and explicated the process by which social capital facilitated subjective well-being, which is the biggest contribution of the present study. First, we extended the current literature on bonding and bridging social capital by showing how they are differently related to subjective well-being. Although prior research found inherent differences between bonding and bridging social capital (e.g., Granovetter, 1973), we are the first to demonstrate that they exhibit different outcomes on subjective well-being. Thus, our finding highlights the importance of social network to understanding social capital and personal well-being, which future research can benefit from.

Second, we found that trait self-esteem played a key role in subjective well-being. Trait self-esteem accounted for nearly two thirds of the total effect between bonding social capital and subjective well-being, and moderated the relationship between bridging social capital and subjective well-being. Hence, although externally driven social capital can contribute to subjective well-being, the key in this process is internally driven trait self-esteem. Yet this may be a result of the subjective nature of subjective well-being. Therefore, future research should investigate how trait self-esteem is related to other types of well-being.

Finally, we suggested that one reason why bonding and bridging social capital are related differently to subjective well-being is their costs of resource provision, which are related to the strength of social ties. Although cost is a construct in economy, it provides a valuable approach to understanding social capital and the dynamics of personal relationships. For instance, consider that the balance between cost and benefit is an important drive of relationship maintenance. Therefore, the cost of providing different types of social capital can be a valuable approach to explaining the exchange of social support and relational satisfaction, which future research can test.

Future Directions

Future research can benefit from the following directions. First, longitudinal research is recommended because this technique allows researchers to detect changes about individual access to social capital, subjective well-being, and trait self-esteem over time and make causal arguments. Second, we conceptualized social capital as perceptions of access to resources in this study. Future research can investigate how enacted social capital—the resources that individuals have received—is related to subjective well-being. Finally, we chose the subjective approach to defining well-being, but there are other types of well-being such as psychological, social, and objective. Future research should investigate how social capital and trait self-esteem influence different types of well-being.

REFERENCE

Aiken, L. S., & West, S. G. (1991). *Multiple regression: Testing and interpreting interactions.* Thousand Oaks, CA: Sage.

Alkire, S. (2002). Dimensions of human development. *World Development, 30*(2), 181–205. doi: 10.1016/S0305-750X(01)00109-7

American Foundation for Suicide Prevention. (n.d.). *Facts and figures*. Retrieved from http://www.afsp.org/understanding-suicide/facts-and-figures

Argyle, M., & Lu, L. (1990). The happiness of extraverts. *Personality and Individual Differences, 11*(10), 1011–1017. doi: 10.1016/0191–8869(90)90128-E

Bishop, J. A., & Inderbitzen, H. M. (1995). Peer acceptance and friendship: An investigation of their relation to trait self-esteem. *The Journal of Early Adolescence, 15*(4), 476–489. doi: 10.1177/0272431695015004005

Bourdieu, P., & Wacquant, L. (1992). *An invitation to reflexive sociology*. Chicago: University of Chicago Press.

Costa, P. T., & McCrae, R. R. (1980). Influence of extraversion and neuroticism on subjective well-being: Happy and unhappy people. *Journal of Personality and Social Psychology, 38*(4), 668–678. doi: 10.1037//0022–3514.38.4.668

DeNeve, K. M., & Cooper, H. (1998). The happy personality: A meta-analysis of 137 personality traits and subjective well-being. *Psychological Bulletin, 124*(2), 197–229. doi: 10.1037//0033–2909.124.2.197

Diener, E., Emmons, R. A., Larsen, R. J., & Griffin, S. (1985). The satisfaction with life scale. *Journal of Personality Assessment, 49*(1), 71–75. doi: 10.1207/s15327752jpa4901_13

Diener, E., Larsen, R. J., & Emmons, R. A. (1984). Person X Situation interactions: Choice of situations and congruence response models. *Journal of Personality and Social Psychology, 47*(3), 580–592.

Diener, E., & Lucas, R. E. (1999). Personality and subjective well-being. In D. Kahneman, E. Diener, & N. Schwarz (Eds.), *Well-being: The foundations of hedonic psychology* (pp. 213–229). New York: Russell Sage Foundation.

Diener, E., Sapyta, J. J., & Suh, E. (1998). Subjective well-being is essential to well-being. *Psychological Inquiry, 9*(1), 33–37. doi: 10.1207/s15327965pli0901_3

Diener, E., Suh, E. M., Lucas, R. E., & Smith, H. L. (1999). Subjective well-being: Three decades of progress. *Psychological Bulletin, 125*(2), 276–302. doi: 10.1037//0033–2909.125.2.276

Diener, E., Wirtz, D., Tov, W., Kim-Prieto, C., Choi, D.-W., Oishi, S., & Biswas-Diener, R. (2010). New well-being measures: Short scales to assess flourishing and positive and negative feelings. *Social Indicators Research, 97*(2), 143–156. doi: 10.1007/s11205–009–9493-y

Dowrick, S. (2004). Income-based measures of average well-being. *World Institute for Development Economic Research*. Retrieved from http://www.wider.unu.edu/stc/repec/pdfs/rp2004/rp2004–24.pdf

Dunning, D., Leuenberger, A., & Sherman, D. A. (1995). A new look at motivated inference: Are self-serving theories of success a product of motivational forces? *Journal of Personality and Social Psychology, 69*(1), 58–68. doi: 10.1037//0022–3514.69.1.58

Ellison, N., Steinfield, C., & Lampe, C. (2011). Connection strategies: Social capital implications of Facebook-enabled communication practices. *New Media & Society, 13*(6), 873–892. doi: 10.1177/1461444810385389

Fitch, G. (1970). Effects of trait self-esteem, perceived performance, and choice on causal attributions. *Journal of Personality and Social Psychology, 16*(2), 311–315. doi: 10.1037/h0029847

Fox, K. R. (1997). The physical self and processes in trait self-esteem development. In K. R. Fox (Ed.), *The physical self: From motivation to well being* (pp. 111–139). Champaign, IL: Human Kinetics.

Furnham, A., & Cheng, H. (2000). Perceived parental behaviour, self-esteem and happiness. *Social Psychiatry and Psychiatric Epidemiology, 35*(10), 463–470. doi: 10.1007/s001270050265

Granovetter, M. (1973). The strength of weak ties. *The American Journal of Sociology, 78*(6), 1360–1380. doi: 10.1086/225469

Granovetter, M. (1983). The strength of weak ties: A network theory revisited. *Sociological Theory, 1*, 201–233. doi:10.2307/202051

Gundelach, P., & Kreiner, S. (2004). Happiness and life satisfaction in advanced European countries. *Cross-Cultural Research, 38*(4), 359–386. doi: 10.1177/1069397104267483

Haring, M. J., Stock, W. A., & Okun, M. A. (1984). A research synthesis of gender and social class as correlates of subjective well-being. *Human Relations, 37*(8), 645–657. doi: 10.1177/001872678403700805

Hayes, A. F. (2013). *Introduction to mediation, moderation, and conditional process analysis: A regression-based approach.* New York: Guilford.

Helliwell, J. F. (2007). Well-being and social capital: Does suicide pose a puzzle? *Social Indicators Research, 81*(3), 455–496. doi: 10.1007/s11205–006–0022-y

Helliwell, J. F., & Putnam, R. D. (2004). The social context of well-being. *Philosophical Transactions of the Royal Society of London B Biological Sciences, 359*(1449), 1435–1446. doi: 10.1098/rstb.2004.1522

Kahneman, D., Diener, E., & Schwarz, N. (1999). *Well-being: The foundations of hedonic psychology.* New York: Russell Sage Foundation.

Ko, H., & Kuo, F. (2009). Can blogging enhance subjective well-being through self-disclosure? *CyberPsychology & Behavior, 12*(1), 75–79. doi: 10.1089/cpb.2008.0163

Kwan, V. S. Y., Bond, M. H., & Singelis, T. M. (1997). Pancultural explanations for life satisfaction: Adding relationship harmony to trait self-esteem. *Journal of Personality and Social Psychology, 73*(5), 1038–1051. doi: 10.1037//0022-3514.73.5.1038

Lin, N. (1982). Social resources and instrumental action. In P. V. Marsden, & N. Lin, (Eds.), *Social structure and network analysis* (pp. 131–145). Beverly Hills, CA: Sage.

Lyubomirsky, S., Tkach, C., & Dimatteo, M. R. (2006). What are the differences between happiness and trait self-esteem? *Social Indicators Research, 78*, 363–404. doi: 10.1007/s11205–005–0213-y

McGillivray, M. (2007). Human well-being: Issues, concepts and measures. In M. McGillivray (Ed.), *Human well-being: Concept and measurement* (pp. 1–22). New York: Palgrave.

Myers, D. G. (2000). The funds, friends, and faith of happy people. *American Psychologist, 55*(1), 56–67. doi: 10.1037//0003-066X.55.1.56

Onyx, J., & Bullen, P. (2000). Measuring social capital in five communities. *The Journal of Applied Behavioral Science, 36*(1), 23–42. doi: 10.1177/0021886300361002

Pavot, W., & Diener, E. (1993). Review of the satisfaction with life scale. *Psychological Assessment, 5*(2), 164–172. doi: 10.1037//1040-3590.5.2.164

Putnam, R. (2000). *Bowling alone: The collapse and revival of American community.* New York: Simon & Schuster.

Resnick, P. (2002). Beyond bowling together: Sociotechnical capital. In J. Carroll (Ed.), *Human-computer interaction in the new millennium* (pp. 247–272). New York: Addison-Wesley.

Rosenberg, M. (1989). *Society and the adolescent self-image* (Rev. ed.). Middletown, CT: Wesleyan University Press.

Ryan, R. M. & Deci, E. L. (2001). On happiness and human potentials: A review of research on hedonic and eudaimonic well-being. *Annual Review of Psychology, 52*, 141–166.

Ryan, R. M., Stiller, J. D., & Lynch, J. H. (1994). Representations of relationships to teachers, parents, and friends as predictors of academic motivation and self-esteem. *Journal of Early Adolescence, 14*(2), 226–249. doi: 10.1177/027243169401400207

Schwarz, N., & Strack, F. (1999). Reports of subjective well-being: Judgmental processes and their methodological implications. In D. Kahneman, E. Diener, & N. Schwarz (Eds.), *Well-being: The foundations of hedonic psychology* (pp. 61–84). New York: Russell Sage Foundation.

Stefanone, M. A., Kwon, K. H., & Lackaff, D. (2012). Exploring the relationship between perceptions of social capital and enacted support online. *Journal of Computer-Mediated Communication, 17,* 451–466.

Steyer, R., Schmitt, M., & Eid, M. (1999). Latent state-trait theory and research in personality and individual differences. *European Journal of Personality, 13*(5), 389–408. doi: 10.1002/(SICI)1099–0984(199909/10)13:5%3C389::AID-PER361%3E3.3.CO;2-1

Symister, P., & Friend, R. (2003). The influence of social support and problematic support on optimism and depression in chronic illness: A prospective study evaluating self-esteem as a mediator. *Health Psychology, 22*(2), 123–129. doi: 10.1037/0278–6133.22.2.123

Tennen, H., & Herzberger, S. (1987). Depression, trait self-esteem, and the absence of self-protective attributional biases. *Journal of Personality and Social Psychology, 52*(1), 72–80. doi: 10.1037/0022–3514.52.1.72

Valkenburg, P. M., Peter, J., & Schouten, A. P. (2006). Friend networking sites and their relationship to adolescents' well-being and social self-esteem. *CyberPsychology & Behavior, 9*(5), 584–590. doi: 10.1089/cpb.2006.9.584

Watson, D., Clark, L. A., & Tellegen, A. (1988). Development and validation of brief measures of positive and negative affect: The PANAs scales. *Journal of Personality and Social Psychology, 54*(6), 1063–1070. doi: 10.1037//0022–3514.54.6.1063

Wellman, B., & Wortley, S. (1990). Different strokes from different folks: Community ties and social support. *American Journal of Sociology, 96*(3), 558–588. doi: 10.1086/229572

Williams, D. (2006). On and off the 'net: Scales for social capital in an online era. *Journal of Computer-Mediated Communication, 11*(2), 593–628. doi: 10.1111/j.1083–6101.2006.00029.x

Yip, W., Subramanian, S. V., Mitchell, A. D., Lee, D. T. S., Wang, J., & Kawachi, I. (2007). Does social capital enhance health and well-being? Evidence from rural China. *Social Science & Medicine, 64*(1), 35–49. doi: 10.1016/j.socscimed.2006.08.027

CHAPTER NINE

Modeling Communication in a Research Network

Implications for the Good Networked Life[1]

DIANA MOK
UNIVERSITY OF WESTERN ONTARIO, CANADA
BARRY WELLMAN AND DIMITRINA DIMITROVA
UNIVERSITY OF TORONTO, CANADA

What happens when "the good life" becomes "the networked life"? For millennia, thinkers thought they knew what the good life was: nestled in a rural or urban village, holding a single stable job preferably in supportive communion with coworkers. Although celebrated in many ways in many centuries, perhaps Thornton Wilder's 1938 *Our Town* epitomized it in the 20th century: life, love, work, family, and death all in a simple village. We note that this pastoral ideal was more a sardonic myth than reality for many peasants, servants, and laborers; nevertheless, the myth still dominated (Marx, 1964). But the "triple revolution" (Rainie & Wellman, 2012) has upset the reality of pastoral village life for many—and perhaps even the myth.

1. The *social network revolution*, starting at least as far back as the 1960s, has seen people change from being embedded in groups—family, community, and work—to involvement in multiple, partial networks.
2. The *Internet revolution* has created communication and information-gathering capacities that dwarf those of the past. Networked computers easily afford connectivity that leaps large distances at a single keystroke. These are *personal* computers: the individual—not the family, community or work group—has become the point of contact. It took the proliferation of the Internet to move to more networked modes of connectivity.

3. The *mobile revolution* has allowed digital media to afford handy access to people and information. Physical separation by time and space can be less important than in olden times.

Our interest here is how the triple revolution has affected how people work together, for it has fostered "networked individualism" where people function more as connected individuals and less as lone workers or embedded group members (Rainie & Wellman, 2012). Until recently, bureaucratic organizations have been the dominant form of North American work. Now, the triple revolution makes it possible for networked workers to move among multiple teams with different interests, enabling workers to connect with nearby and distant network members by digital media as well as in person. Many are also spatially dispersed "distributed workers," using digital media to communicate and exchange information, and using planes and cars for in-person contact.

It is not just the workers, but organizations too, that have become networked. Over time, the triple revolution has fostered changes in the broader social context that have affected organizations' structure, authority relations, and information flows (Olson & Olson, 2013). Networked organizational forms have gained traction as labor has moved from "atom work"—growing, mining, making, and transporting things—to "bit work"—selling, describing, and analyzing things (Florida, 2002; Negroponte, 1996; Rainie & Wellman, 2012). In contrast to traditional bureaucratic organizations, networked organizations often have flexible boundaries between groups, and individual workers are affiliated with multiple groups.

While early enthusiasts celebrated networked organizations as fostering the good life that liberated workers from stultifying bureaucracies (e.g., Grantham & Nichols, 1993), research is now providing more complex descriptions showing how networked work often nestles within formal hierarchies (Heckscher, 1994; Krebs, 2007; Stephenson, 2008). In such organizations, many workers are no longer members of a single work group but juggle involvement in a multitude of often-intersecting projects and teams (Dimitrova & Wellman, 2015). For them, the good working life is no longer a comfortable, settled job with reasonable pay; the good life is the ability to juggle relationships and responsibilities without being captured by any one of them.

NETWORKED SCHOLARS: THE GRAND NETWORK

For the past few years, we have been investigating a special kind of networked workers: networked scholars. We have been wondering if networked life is a good life for scholarly researchers. When scholars become involved in research networks, they lose the cloistered life of sitting alone in their studies scribbling on their foolscap or computers (Deresiewicz, 2014), and they lose the bureaucratic certitude of being ensconced in a siloed discipline or a single project (Jacobs, 2013).

By contrast, scholars in research networks are often connected to people in multiple disciplines and dispersed across countries and continents. Neither isolated nor siloed, they usually have partial involvements in multiple projects. Yet such networks are often in flux. Network scholars have much scope about who to connect with and in what roles. To what extent is this "the good scholarly life"? It is this aspect of networked work that we address in this chapter.

Our research focus has been on a large scholarly network: the Graphics, New Media and Design Network of Centres of Excellence (GRAND NCE), a Canada-wide, multi-disciplinary, and multi-institutional network of researchers that started in 2010 (for detailed descriptions, see Dimitrova, Mok, & Wellman, 2015; Mo & Wellman, 2014). The GRAND network stretches 5,000 kilometers from Victoria on the Pacific Coast to Halifax on the Atlantic coast, with most ties being in the largest English-speaking provinces: Alberta, British Columbia, and Ontario (see Figure 9.1). Nearly half of the researchers (46%) are computer scientists, while most of the rest are researchers in media and/or information studies and other social scientists. As GRAND is a start-up that we have been studying over time, we are in a good position to see how the researchers' networks change as the network matures.

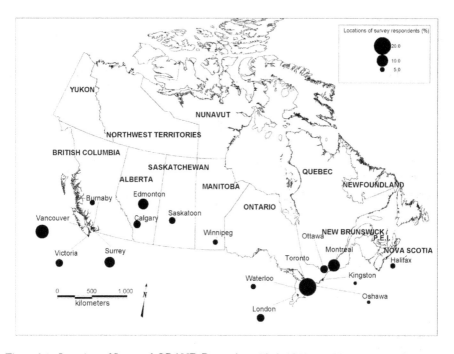

Figure 9.1. Location of Surveyed GRAND Researchers, 2010–2012, N = 83.

Our primary sources for this chapter are two online surveys we conducted with GRAND faculty researchers: in 2010, at the initial stage of the network, and in 2012, at its mid-point. In each survey, we asked the researchers which GRAND members they KNOW; which they are FRIENDS with; who they give or get ADVICE from; and who they WORK with in writing papers, giving presentations, and developing applications. We created three related datasets for each of the four networks derived from the researchers' responses:

1. The 2010 NAVEL survey dataset (N = 144), with a 70% response rate. The resulting networks are defined on 143 (N-1) GRAND faculty researchers, forming a network database of 83 x 143 = 11,869 ties.
2. The 2012 NAVEL survey dataset (N = 207), with a 60% response rate. The networks are defined on 206 (N-1) GRAND faculty researchers, forming a network database of 83 x 206 = 17,098 ties.
3. Repeat respondents (N = 83) comprises researchers who completed both the 2010 and 2012 surveys. *This is the dataset we principally use*: 57% of the original GRAND researchers and 82% of those who responded to the initial 2010 survey. It does not take into account the small number of researchers who left GRAND between 2010 and 2012 and the somewhat larger number who joined between 2010 and 2012.

All of these self-identified networks are asymmetrical, as they encompass the frequent situation of researchers saying they have relationships with specific others, but those others not reciprocating the relationship. We also use for background information our semi-structured interviews with 47 GRAND researchers, as well as our participant-observation in four projects and at all GRAND annual meetings.

In our analysis, *disciplines* refer to eight broad disciplinary categories defined by the major Canadian funding agencies. *Senior* versus *junior* researchers refers to tenured versus non-tenured professors. *PNIs* and *CNIs* stand for Principal and Collaborative Network Investigators, designated roles in GRAND. PNIs receive more funding from the network, participate in more projects, and typically lead at least one project. CNIs receive less funding, participate in one or only a few projects, and do not lead projects. The division does not follow academic seniority: While some CNIs are non-tenured academics invited by former supervisors to the network, some are tenured.

Studying this research network involves multiple challenges, as we cannot simply study networked scholars as disconnected aggregates of individuals—the standard statistical fare of regressions, et al.—nor as inhabitants of single bureaucratic cocoons. Instead, we use three models to understand the communication networks of these scholars from multiple disciplines interacting in multiple research projects.

Stability and Change Model

GRAND researchers started with networks that were cross-disciplinary, geographically dispersed, and cross-institutional, and for the most part, their networks remained stable over the two-year period. Although GRAND's membership grew, most individual researchers' own KNOW, FRIEND, ADVICE, and WORK networks within GRAND did not grow as rapidly as GRAND membership did. Overall, most researchers changed no more than 4 to 8 of their ties out of a mean of 9 to 23 in each of their four networks (Figure 9.2). *Changed ties* refer to ties being added or dropped in each network, as contrasted with no change. Those ties that changed were in the researchers' ADVICE, FRIEND and WORK ties. The KNOW network ties grew as the researchers needed little effort to maintain them—many connected only at the annual GRAND meeting or other conferences. WORK networks shrank in size and decreased in geographical diversity. After two and a half years, researchers knew more people from different disciplines, institutions, and cities but collaborated with fewer colleagues and based their collaboration to a greater extent locally.

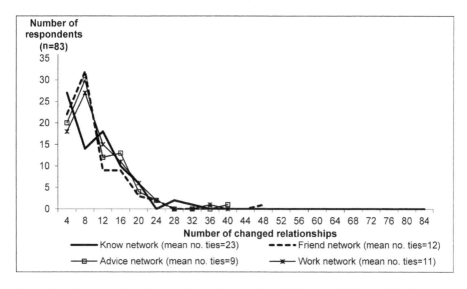

Figure 9.2. Frequency Distribution of Survey Respondents by Number of Changed Ties in 2010–2012.

Dynamic Model

Although most relationships were stable, what about those relationships that did change? We developed a model to help study networks with changing membership and relationships. The model helps investigate which attributes of the researchers have exerted stronger impact on maintaining their four types of relationships: When do they drop existing ties, add new ties, keep old ties, or do nothing (the null choice)?

We analyze the 83 GRAND members who completed both the 2010 and 2012 surveys and their existing or potential relationships with the 143 researchers who were members of GRAND in both 2010 and 2012. In our multilevel multinomial logistic regression model, we recognize both the network structure of the dataset and the polytomous responses of the categories. We estimate four separate multilevel multinomial logistic models, for each type of tie.

To begin, let m be the number of unordered response categories (such as presence of a tie) and let the linear predictors be denoted by

$$\eta_{ij}^{(m)} = \alpha^{(m)} + \beta^{(m)} x_{ij} + \xi_j^{(m)} + \delta_{ij}^{(m)} . \tag{1}$$

The multinomial logistic link is defined as

$$P\left(Y_{ij} = m \mid x_{ij}, \xi_j, \delta_{ij}\right) = \frac{\exp\left(\eta_{ij}^{(m)}\right)}{1 + \sum_{k=2}^{4} \exp\left(\eta_{ij}^{(k)}\right)} . \tag{2}$$

Note that in both equations (1) and (2), the subscript $j = 1,..,83$ denotes the 83 repeat survey respondents, and the subscript $i = 1,...,143$ indexes the other 143 GRAND members in 2010, who are the existing or potential network ties. In this model specification, we assume that there are no category-specific covariates in x_{ij}. Each of the m-1 equations in (1) has category-specific parameter $\alpha^{(m)}$ and $\beta^{(m)}$. The entities $\xi_j^{(m)}$ and $\delta_{ij}^{(m)}$ are vectors of random errors, representing unobserved heterogeneity at the respondent (j) and the tie (i) level, respectively. In the error component of the model, we assume that $\xi_j^{(m)}$ and $\delta_{ij}^{(m)}$ are independent and that all $\xi_j^{(m)}$ and $\delta_{ij}^{(m)}$ are independent and identically distributed as normal, centered at a mean of zero with a variance-covariance matrix of Σ_ξ and Σ_δ, respectively. The residual variance at the tie level is not identified: It can be set to any arbitrary value (Skrondal & Rabe-Hesketh, 2003). For simplicity, Σ_δ is set to be $\frac{1}{m}(I+M)$, where I and M are m-1 dimensional identity and unit matrices, respectively.

The odds of any two response categories t and s are:

$$\frac{P\left(Y_{ij} = t \mid x_{ij}, \xi_j, \delta_{ij}\right)}{P\left(Y_{ij} = s \mid x_{ij}, \xi_j, \delta_{ij}\right)} = \exp\left(\eta_{ij}^{(t)} - \eta_{ij}^{(s)}\right) , \tag{3}$$

implying the well-known assumption of Independence of Irrelevant Alternatives (IIA). However, since IIA does not hold marginally with respect to the random errors, the inclusion of random terms in the linear predictors allows the model to partially relax the IIA assumption.

All four polytomous variables (KNOW, FRIEND, ADVICE, and WORK) have four unordered categorical responses: The variable receives a "1" if the researchers have kept their 2010 tie in 2012 (the reference category). It receives a "2" if the researchers have dropped the tie by 2012; a "3" if the researchers have formed a new tie in 2012, and a "4" if the researchers have never formed a tie with a GRAND network member between 2010 and 2012 (the null category).

The covariates are all dichotomous. They include variables that describe the characteristics of each of the 83-respondent-to-143-potential (existing)-tie relationship, focusing on cross-disciplinary, cross-institutional, cross-city, or cross-provincial relationships. They also identify relationships formed by researchers of different genders or different professorial ranks. In addition, two dichotomous variables have been created to identify the characteristic of the researchers: Whether the researcher is a central (PNI) or peripheral (CNI) network member, and whether the researcher is an assistant or tenured professor. All the models are estimated using MCMC routines implemented in the MCMCglmm package in R.

The results of the multilevel multinomial logistic regression model are summarized in Table 9.1. The coefficient estimates are interpreted as the impact of changing the covariates on the log-odds of choosing a particular choice category, relative to the reference choice of keeping the existing tie. The fact that these impacts are logs of relative probabilities makes them tricky to interpret as their signs might not be the same as the covariates' actual marginal impacts on the probabilities of the choices. Elsewhere, we have summarized the marginal impacts of the covariates (Dimitrova et al., 2015); here, we report the actual parameter estimates.

Table 9.1. Multilevel Multinomial Logit Model Results: KNOW, FRIEND, ADVICE, and WORK Networks, 2010–2012.

From Relationship in 2010	KNOW			FRIEND		
	Drop	Add	Null	Drop	Add	Null
Characteristics of relationship						
Cross-discipline	0.5893 **	0.4893 **	1.2878 **	0.4705 **	0.3575 **	1.6911 **
Cross-institution	0.8620 **	0.6586	1.4501 **	0.2684	1.4454 **	2.2813 **
Cross-city	0.6349 **	0.1887	0.6522 **	0.8206 **	-1.3254 **	0.5008 **
Cross-province	0.2896	0.7102	1.3832 **	0.4518	-0.6949 **	1.0392 **
Different gender	0.0256	-0.2522 **	0.0634	-0.1173	0.0777	0.2225 **
Cross-professorial rank	0.0805	-0.0715	0.1443	-0.0175	0.1799	0.3473 **

From Relationship in 2010	KNOW			FRIEND		
	Drop	Add	Null	Drop	Add	Null
Characteristic of respondent						
CNI in 2010	0.5582 *	-0.3650 *	0.7985 **	0.4273 *	0.4273 *	0.6719 **
Junior	-0.1075 *	0.1331 *	-0.0712 *	-0.5469 *	-0.5469 *	0.1767 *
Intercept	-4.0455 **	-4.2568 **	-3.2692 **	-0.9408 **	-3.0138 **	-1.5648 **
Predicted probability at sample means	0.0361	0.0144	0.2988	0.0336	0.0035	0.0285
	ADVICE			WORK		
	Drop	Add	Null	Drop	Add	Null
Characteristics of relationship						
Cross-discipline	-0.0489	0.2386	1.4317 **	-0.0609	0.5800 **	1.4526 **
Cross-institution	0.6160 *	0.7210 *	1.6438 **	0.3689	0.6994	1.8224 **
Cross-city	0.2810	-0.1605	0.9263 **	0.4792 *	-0.0676	0.8891 **
Cross-province	0.4111	-0.2108	1.0076 **	0.5105 *	0.4527	1.2624 **
Different gender	-0.0174	-0.2330	0.0294	0.0306	0.0516	-0.0444
Cross-professorial rank	-0.1831	0.0280	0.1639	0.1490	0.0482	-0.0054
Characteristic of respondent						
CNI in 2010	0.5592 *	-0.1595 *	0.7364 **	0.3536 *	-0.1480 *	0.7180 **
Junior	-0.1805 *	0.3131 *	0.0103 *	-0.5283 *	0.7275 *	0.2278 *
Intercept	-0.7088 **	-1.5903 **	-0.4370	0.2597	-2.9529 **	-0.2685
Predicted probability at sample means	0.0334	0.0090	0.0242	0.0506	0.0040	0.0155

Notes: "Drop" refers to dropping an existing tie, "Add" refers to adding a new tie, and "Null" refers to maintaining no connection with a GRAND member in both 2010 and 2012—the null choice. Predicted probabilities are evaluated at sample means. Double asterisks identify estimates that are significant at the 5% level; single asterisks, the 10% level. The left-out default category is keeping the tie that has existed since 2010 ("Keep"). All estimates are interpreted as the impact of the covariates on the log-odds of Drop, Add, or Null, relative to Keep, based on 83 respondents and 143 potential ties.

We summarize our results about what drives ties to be added or dropped in five points (for more details, see Dimitrova et al., 2015):

1. For the KNOW and FRIEND networks, diversity—disciplinary, geographic, institutional—increases the log-odds of relationships being dropped. Although the researchers know more people from different provinces and befriend more researchers from different institutions, they are also just as likely to drop such relationships.

2. For ADVICE and WORK networks, each diversity dimension has a somewhat different impact. While multidisciplinary diversity increases the log-odds of ADVICE relationships being added and WORK relationships being kept, geographic and institutional diversity increases the log-odds of ties being dropped.

3. Ties between men and women or between junior and senior faculty do not matter with respect to ties being kept or changed: Their coefficient estimates are mostly insignificant. Network changes are driven by disciplinary, geographic, and institutional diversity rather than the personal attribute of gender or the university status of being tenured.

4. However, university status does affect how the networks of non-tenured professors change. They are more likely than tenured professors to keep or add ties to their KNOW, ADVICE, and WORK networks.

5. The more peripheral CNIs are more likely than the more central PNIs to drop their KNOW, FRIEND, ADVICE, and WORK ties.

Although the model estimates are small in magnitude, most of them are statistically significant and are, more importantly, contextually meaningful: For example, a one percent increase (a coefficient estimate of the order of 0.01) implies the probability of adding one or two collaborator(s) out of 143 network members to a personal work network. While the number "one percent" might seem small, the contextual meaning of two collaborators is meaningful, considering the fact that the mean number of work ties is about 11 in 2010 and 7 in 2012.

Transitions Model

Which kinds of relationships among GRAND researchers changed in the two-year period—that is, where did the growth and decline of the KNOW, FRIEND, ADVICE, and WORK networks actually come from? We show the transition from one type of relationship to another (such as ADVICE to WORK) between 2010 and 2012, based on the 83 respondents in both surveys and the 143 potential 2010 network ties. Tie categories shown in Table 9.2 are mutually exclusive; the null category means no ties were formed. The row and column sums show the total number of ties in each of the 16 categories in 2010 and 2012, respectively.

Table 9.2. Transition Matrix of KNOW, FRIEND, ADVICE, WORK, and Cross-Tabulated Networks, 2010–2012.

2010	2012 KNOW only	KNOW and FRIEND	KNOW and ADVICE	KNOW and WORK	KNOW and FRIEND and ADVICE	KNOW and FRIEND and WORK	KNOW and ADVICE and WORK	KNOW and FRIEND and ADVICE and WORK	NULL	ROW SUM
KNOW										
KNOW only	372	26	48	10	26	4	7	9	146	648
KNOW and FRIEND	121	62	9	6	27	8	2	7	33	275
KNOW and ADVICE	16	2	7	2	3		2	6	2	40
KNOW and WORK	53	5	23	6	16	3	12	12	22	152
KNOW and FRIEND and ADVICE	19	17	3	1	16	3	4	10	2	75
KNOW and FRIEND and WORK	35	21	12	4	24	2		14	11	123
KNOW and ADVICE and WORK	39	1	32	11	13	6	12	24	11	149
KNOW and FRIEND and ADVICE and WORK	66	68	17	11	78	40	29	134	15	458

Transition matrix of network ties (rows = 2010 status, columns = 2012 status):

2010 \ 2012	KNOW only	FRIEND only	FRIEND and ADVICE	FRIEND and WORK	FRIEND and ADVICE and WORK	ADVICE only	ADVICE and WORK	WORK only	NULL	ROW SUM
KNOW only	372	26				48				648
FRIEND only	3	1	1		4				11	20
FRIEND and ADVICE						1				1
FRIEND and WORK	2								7	9
FRIEND and ADVICE and WORK					1		1		3	5
ADVICE only		2				3			5	10
ADVICE and WORK	4	1	1	2	1	4			6	19
WORK only	5								6	11
NULL	255	13	53	31	19	21	3	11	9468	9874
COLUMN SUM	988	218	208	83	228	95	70	231	9748	11869
GROWTH/DECLINE in 2012	340	-57	168	-69	153	-54	-53	-227	-126	75

Notes: The transition matrix shows the counts of ties that have changed from one type of network to another between 2010 and 2012—for example, of the 648 KNOW-only ties in 2010 (row sum of KNOW-only ties in 2010), 372 stayed as KNOW-only in 2012, but 26 became FRIENDS and 48 became ADVICE ties in 2012. Meanwhile, among the 988 KNOW-only ties in 2012 (column sum of KNOW only in 2012), 53 such ties used to be WORK ties in 2010, but the formal work relationship no longer existed in 2012. Growths/declines were calculated based on the difference between the column sum (2012 total) and the row sum (2010 total) in each category. Calculations are based on 83 respondents and 143 potential 2010 network ties.

The KNOW-only relationships had the greatest growth between 2010 and 2012, while ties combining KNOW, FRIEND, ADVICE, and WORK had the greatest decline (see Table 9.3). In both instances, much of the growth and decline was attributable to researchers redefining their 2010 relationships into different categories in 2012.

The largest growth between 2010 and 2012 was the transformation of ties into KNOW-only relationships, rather than the combination of KNOW and some other relationship such as ADVICE. Thus, much of the growth in KNOW-only ties was attributed to redefining former relationships instead of forging new ties that did not exist in 2010. The researchers kept their KNOW acquaintance-ships, but cut down on their more intense FRIEND, ADVICE, and WORK relationships.

In 2010, 648 ties were KNOW-only, but 988 such cases were reported in 2012, representing a net growth of 340 KNOW-only cases. Of these 340 ties, less than a third was from forging new ties that did not exist in 2010 (n = 109); more than two-thirds (n = 231) were based on redefining existing ties. In fact, the largest growth occurred from redefining former KNOW-FRIEND relationships into KNOW-only relationships, with a net gain of 95 ties into KNOW-only from the KNOW-FRIEND category (see Table 9.3). Put differently, 95 KNOW-FRIEND relationships dropped their friendship. Meanwhile, the second largest growth occurred when 57 of the combined KNOW-FRIEND-ADVICE-WORK ties dropped all other social and formal relationships and became acquaintances in 2012.

KNOW-ADVICE relationships grew by a net of 168 ties between 2010 (n = 40) and 2012 (n = 208); like KNOW-only, much of its growth came from researchers redefining former relationships. In 2010, there were only 40 (2% of 1,995 existing ties) such ties; in 2012, it grew to 208 cases (10% of 2,121 existing ties), a net gain of 168. Of the 168 additional cases reported in 2012, 51 cases were formed from forging new KNOW-ADVICE ties that did not exist in 2010. The remaining 117 cases were based on researchers redefining former relationships, with 32 cases formed from formerly KNOW-only relationships (gaining advice), 30 cases from formerly KNOW-ADVICE-WORK (dropping work relationships), and 21 cases from formerly KNOW-WORK (changing from work to advice relationships).

The KNOW-FRIEND-ADVICE category also grew in number, by a net of 153 ties from 75 ties in 2010 to 228 ties in 2012. Yet, of the total net growth, only 17 ties were new ties while 136 were from redefining former but different relationships. In many cases the growth in this category came from a reduction in the different types of ties two researchers had.

Most prominently, as the initial excitement about GRAND membership wore off and the quondam pressure of scholarly work kicked in, WORK relationships

Table 9.3. Net Gains and Losses in Various Types of Relationships, 2010-2012.

Types of former relationships	2012 relationships								
	KNOW only	KNOW and FRIEND	KNOW and ADVICE	KNOW and WORK	KNOW and FRIEND and ADVICE	KNOW and FRIEND and WORK	KNOW and ADVICE and WORK	KNOW and FRIEND and ADVICE and WORK	NULL
KNOW									
KNOW only	0	-95	32	-43	7	-31	-32	-57	-109
KNOW and FRIEND	95	0	7	1	10	-13	1	-61	20
KNOW and ADVICE	-32	-7	0	-21	0	-12	-30	-11	-51
KNOW and WORK	43	-1	21	0	15	-1	1	1	-9
KNOW and FRIEND and ADVICE	-7	-10	0	-15	0	-21	-9	-68	-17
KNOW and FRIEND and WORK	31	13	12	1	21	0	-6	-26	8
KNOW and ADVICE and WORK	32	-1	30	-1	9	6	0	-5	-10
KNOW and FRIEND and ADVICE and WORK	57	61	11	-1	68	26	5	0	4
NULL	109	-20	51	9	17	-8	10	-4	0
Net gains/losses 2010-2012	340	-57	168	-69	153	-53	-54	-227	-126

Notes: To see how we arrive at the numbers, consider first the row KNOW-only in Table 9.2 as an illustration. In 2010, a total of 648 KNOW-only ties were observed (Table 9.2). Of these 648 ties, 26 became KNOW-FRIEND in 2012 (an outflow). Now consider the column KNOW-only in Table 9.2. In 2012, a total of 988 KNOW-only was observed, of which 121 were KNOW-FRIEND in 2010 (an inflow). Comparing 26 outflows and 121 inflows of KNOW-FRIEND, there was a net inflow of 95 KNOW-only from the former KNOW-FRIEND category, as shown in cell row 2 column 1 here.

declined. The greatest decline was in the KNOW-FRIEND-ADVICE-WORK category where WORK or FRIEND relationships were dropped. In 2010, 458 KNOW-FRIEND-ADVICE-WORK ties were reported, but the count dropped by 50% to 231 in 2012. Many (68) were lost to KNOW-FRIEND-ADVICE— that is, researchers simply dropped their work relationships. The second largest loss was to KNOW-FRIEND (n=61)—that is, researchers dropped both their advice and work relationships, but remained as friends. The third largest loss (n = 57) was to KNOW-only; researchers simply lost all their social and formal ties and became acquaintances only.

Ties that were initially KNOW-WORK in 2010 experienced the second largest decline by 2012, a net loss of 69 ties although it was far less in magnitude than the 227 tie decline in the KNOW-FRIEND-ADVICE-WORK combination. Much of the loss was to KNOW-only (n = 43): Researchers became acquaintances and dropped their work relationships. Some KNOW-WORK ties were transformed to become KNOW-ADVICE (n = 21), while some others became friends in addition to providing advice for each other (KNOW-FRIEND-ADVICE n = 15).

In sum, much of the growth in the KNOW network was due to existing relationships being redefined, dropping friendship as well as work or advice relationships. Among the new ties that were formed by 2012, most were new (KNOW) acquaintances, while slightly less than half were new advice relationships. At the same time, a small number (n = 32) of existing FRIEND, WORK, or ADVICE ties were completely dropped by 2012.

CONCLUSIONS

How have GRAND researchers networked? Several results stand out: the relative stability of the network, the volatility of ties, the general direction of the changes from rich multifaceted ties to less complex ties, and the role of diversity as driving the changes.

GRAND is a voluntary body—sweetened by moderate research grants—so the participants came to it willingly—for the adventure, stimulation, excitement of being in a large enterprise, and to be sure, for the funding. At the initial stages, many ties were made, rekindling old relationships, continuing existing partnerships, and forging new ones. GRAND researchers were encouraged to forge ties with scholars in other disciplines and other universities, and for the most part, they followed suit. In our interviews, the researchers told us that the initial stages of the GRAND network were like the start of attending university, as they scrambled to find compatible or advantageous others, and then dropped some that did not work out.

GRAND was—and remains—a heavily networked research enterprise. Nevertheless, the first flush of enthusiasm has faded and relationships have adjusted for the long term. For starters, most researchers' networks became smaller and thinner. They contained few members, and although most ties continued, the average tie had few role relationships within the tie. The most common change was in the hardest relationship to maintain: WORKing together to produce papers, presentations, and computer applications. By contrast, just being acquainted with other GRAND members—what we call KNOWing them—increased in size and diversity. In many cases, the researchers only saw them at the annual meetings or through occasional emails, but it should be noted that most GRAND researchers were excited to be in contact with notable people in their own field, exciting newcomers, and interesting people from heretofore unknown disciplines. "What exactly is a sociologist?" Wellman was sometimes asked.

But this chapter is about change as much as stability. Although GRAND researchers came to know more people from different provinces and befriend more researchers from different institutions, such ties were short-lived at times. Instead of either steadily growing or dropping ties, networked researchers test drive and redefine ties as they see fit thereby making the boundaries of their personal networks permeable. While ties are usually expected to become more complex over time, the prevalent change in GRAND is from broad to narrower relationships. The dynamics of deliberately created networks might be different from naturally occurring ones.

Diversity—pro and con—plays an important role in the dynamic of change. While GRAND had a recruiting bias toward diversity, in practice, disciplinary, geographic, and institutional diversity decreased the chance of ties being kept and increased the chance of ties being dropped in the KNOW and FRIEND networks. There is a comfort zone in homophily—speaking the same disciplinary language, meeting at the same conferences, and applying to the same granting agencies. (The Canadian government sharply discriminates between computer researchers and social science/humanities researchers in a much more poorly funded agency.) WORK networks were especially apt to shrink in size and to lose researchers at more distant universities.

Yet, there is a delicate balance between diversity and similarity in GRAND. Where similarity gives cultural comfort and pre-existing ways of working, diversity provides fresh ideas and adventures. That is why the ostensibly weak tie of KNOWing someone else is important—they are more apt to come from different places and meet you there. Diverse ADVICE and FRIENDship ties also open up new ways of seeing.

Each dimension of diversity has had a somewhat different impact for the ADVICE and WORK networks. Multidisciplinarity increases the likelihood of relationships being added (or being kept in the case of WORK networks), but geographic and institutional diversity increases the probability of ties being dropped. Adding and dropping relationships occur together, affected by different types of diversity.

In the GRAND network, some researchers are more central than others. PNIs belong to more projects, and are more like to keep all kinds of relationships: KNOW, FRIEND, ADVICE, and WORK. By contrast, CNIs are more likely to come and go, and when they stay, they are more likely than Principals to trim down their ties by limiting their role relationships. By contrast, non-tenured faculty members, be they PNIs or CNIs, are more likely than tenured professors to keep or add ties to their KNOW, FRIEND, ADVICE, and WORK networks. As savvy young faculty members, they are hungry to add productive ties to their usually-smaller networks.

Ties between researchers with different genders have little impact on changes in their networks. There are more men than women in GRAND to be sure, but GRAND's focus on human-computer interaction is a computer science field with a comparative high percentage of women, and many women have senior positions in the GRAND network. It is the research ties that affect GRAND network dynamics, not the gender of the participants.

GRAND researchers have become jugglers—or perhaps a better metaphor is whirling dervishes—moving among projects and relationships while adding and dropping ties in the process. Tie multiplexity is coupled with tie volatility. Diversity reminds us that perhaps sometimes there might be too much of a good thing. When diversity becomes too much to handle, researchers fall back to more homophilous relationships. The triple revolution has gone to work, and has produced a hyperconnected version of "the good life" for many of the networked scholars we have studied.

NOTE

1. Acknowledgments: NAVEL participants Christian Beermann, Anatoliy Gruzd, Tsahi Hayat, and Guang Ying Mo have been an integral part of this project. We thank the GRAND Network of Centres of Excellence for its research support and Director Kelly Booth for his good advice and good cheer. We also thank the Faculty of Information at the University of Toronto for being a hospitable base and Dean Seamus Ross for his support.

REFERENCES

Deresiewicz, W. (2014). *Excellent sheep*. New York: Free Press.

Dimitrova, D., Mok, D., & Wellman, B. (2015, forthcoming). Changing ties in a far-flung, multidisciplinary research network. *American Behavioral Scientist, 59*(4).

Dimitrova, D., & Wellman, B. (2015, forthcoming). Networked work and network research. *American Behavioral Scientist, 59*(4).

Florida, R. (2002). *The rise of the creative class.* New York: Basic.

Grantham, C., & Nichols, L. (1993). *The digital workplace.* New York: Van Nostrand Reinhold.

Heckscher, C. (1994). Defining the post-bureaucratic Type. In C. Heckscher & A. Donnellon (Eds.), *The post-bureaucratic organization.* (pp. 14–62). Thousand Oaks, CA: Sage.

Jacobs, J. (2013). *In defense of disciplines.* Chicago: University of Chicago Press.

Krebs, V. (2007). Managing the 21st century organization. *Institute of Human Resources and Information Management, 11*(4), 2–8.

Marx, L. (1964). *The machine in the garden.* New York: Oxford University Press.

Mo, G. Y., & Wellman, B. (2014). Using multiple membership multilevel models to examine networks in networked organizations. *Knowledge Media Design Institute Report, 14–5.*

Negroponte, N. (1996). *Being digital.* New York: Vintage.

Olson, J., & Olson, G. (2013). *Working together apart.* San Rafael, CA: Morgan & Claypool.

Rainie, L., & Wellman, B. (2012). *Networked.* Cambridge, MA: MIT Press.

Skrondal, A., & Rabe-Hesketh, S. (2003). Multilevel logistic regression for polytomous data and rankings. *Psychometrika, 68*(2), 267–287.

Stephenson, K. (2008). *The quantum theory of trust: Power, networks and the secret life of organisations.* Upper Saddle River, NJ: Prentice Hall.

Wilder, T. (1938). *Our town.* New York: Henry Miller's Theater.

CHAPTER TEN

Communicating Online Safety

Protecting Our Good Life on the Net[1]

ROBERT LAROSE, SALEEM ALHABASH, MENGTIAN
JIANG, RUTH SHILLAIR, HSIN-YI SANDY TSAI,
SHELIA R. COTTEN AND NORA J. RIFON
MICHIGAN STATE UNIVERSITY, USA

CONCEPTUALIZING ONLINE SAFETY

A succession of news-making online threats introduce online safety behavior—defined as actions that users take to protect their computing devices and themselves from online security threats—as a topic of interest for communication researchers. These episodes highlight the importance of individual safety behavior and the need to motivate users to take protective measures of their own. For example, the Heartbleed bug discovered in April 2014 forced users of many popular websites using Open SSL to encrypt secure transactions to change their passwords. The following month, hacker attacks orchestrated by the Chinese government that preyed upon employees of major U.S. corporations, who opened innocuous-seeming email attachments purportedly from corporate security officers or other trustworthy sources, were revealed. Each day, millions of "phishing" attacks that target consumers with the help of information gleaned from social media sites succeed in implanting malware on the computers of users (National Cyber Security Alliance, McAfee, & JZ Analytics, 2012) and entrapping many of them in fraud (Martin & Rice, 2013). A fifth of U.S. Internet users still lack basic protections, and a third of users worldwide fail to keep security patches updated for popular computer applications such as Word, Java, and Flash (National Cyber Security Alliance et al., 2012; Seidman, 2012). How can we motivate users to be more effective at protecting themselves against the slew of risks they face every time they use their computers and the Internet?

Online safety protection is somewhat analogous to health communication that aims to protect individuals from threats to their well-being. The analogy was recognized in previous published works by our team (LaRose, Rifon & Enbody, 2008; Lee, LaRose, & Rifon, 2008) based on Protection Motivation Theory (PMT; Rogers, 1975; Rogers & Prentice-Dunn, 1997) that has informed a stream of literature in the Management Information Systems (MIS) domain (e.g., Anderson & Agarwal, 2010; Davinson & Sillence, 2010; Ifinedo, 2012; Lai, Li, & Hsieh, 2012). As with behaviors that preserve one's health, online protection behaviors are conceived to result from appraising threats to well-being and the resources available to cope with the threats, together with the costs and benefits of both the adaptive and maladaptive behaviors. Contagious diseases are a particularly apt analogy since online threats spread through contact with others, as the adaptation of the virus terminology in the online safety realm readily reveals. As with health consumers and their medicines, computer users may be lulled into a false sense of safety by reliance on automated protections, analogous to ceasing the use of hand wipes after getting a flu shot.

The purpose of this chapter is to re-integrate online safety behavior research with the Protection Motivation Theory (PMT) paradigm that spawned it and that has guided a great deal of research in health communication. The chapter starts with consideration of the unique aspect of risk communication in the online safety domain and traces the development of social science theories from the fields of communication, social psychology, and management information systems that have been applied to further understanding of online safety behavior. Next, we review previous research by our team that has not been widely circulated in communication research circles for its contributions to online safety research and the PMT paradigm on which it is based. This is followed by reflections on current practices in online safety communication and public policies that affect those practices. We also examine qualitative research that affords new opportunities for development of risk communication strategies, before concluding with recommendations for future research, improved risk communication practices, and public policies.

Online Safety: Toward an Integrated Theory

While online safety research can be situated in health and risk communication domains, there are some unique aspects of online risks that expose new opportunities for expanding our understanding of risk communication beyond routine applications of well-recognized theories like PMT. Online protection behavior is about protecting an "other" in the form of a computing device and its applications rather than the "self." Thus the threats are one step removed from the usual bounds of health communication, although the health of a child or relative is somewhat analogous (Campis, Prentice-Dunn, & Lyman, 1989). The responsibility for our digital well-being is thus considerably more diffuse than for our physical well-being, even more so because of the many entities involved in maintaining it in addition to our selves: Internet service providers, software and hardware manufacturers, website proprietors, and law enforcement as well as the information technology (IT) professionals and security software designers and manufacturers that are analogous to medical and pharmaceutical providers in the health domain. Also, the protected entity is non-human and expendable. Even though we may endow our computing devices with human qualities (Reeves & Nass, 1996), the fact remains that they are things that can be bought and sold and disposed of and replaced if they become too seriously impaired, or simply go out of style; factors that may affect threat appraisals. Finally, highly organized criminal and state-sponsored enterprises are relentlessly at work to destroy the good life we enjoy online.

The theoretical contributions of online safety research can be integrated with health communication research by returning to the origins of PMT and tracing the evolution of related theories across three decades. Recent developments in health communication, social psychology, and management information systems research contribute to this synthesis, adding new variables and contributing new insights into the nature of the relationships among key concepts.

PMT (Rogers, 1975; Rogers & Prentice-Dunn, 1997) traces its origins to Expectancy-Value Theory (EVT, e.g., Atkinson, 1964; Fishbein & Ajzen, 1975) and Social Cognitive Theory (SCT, Bandura, 1977). From the former, PMT proposed that beliefs about the expected outcomes of behavior (i.e., the vulnerability to a threat such as lung cancer or identity theft) and the evaluation of those beliefs (i.e., the severity of the threat) determine protection motivation, or the intention to engage in adaptive behavior (e.g., smoking cessation or automatic virus protection updates). The benefits of a maladaptive behavior (e.g., the pleasure of cigarette smoking or the convenience of online banking) combine with threat severity and vulnerability to form a threat appraisal. SCT's self-efficacy was the basis for coping self-efficacy in PMT, which is defined as beliefs in one's ability to enact adaptive (i.e., protective) behavior. PMT added response efficacy, or belief in the effectiveness of the adaptive behavior to the coping appraisal, and an evaluation of the costs of the adaptive response (e.g., nicotine withdrawal symptoms, the financial cost of protective software) to comprise coping appraisal. Together, threat and coping appraisals determine protection motivation.

PMT thus parallels two other theories that have also been applied to risk communication research: the Theory of Planned Behavior (TPB, Ajzen, 1991) and Social Cognitive Theory (SCT, Bandura, 1986). TPB also combines an expectancy-value formulation (employed there to define attitudes) and a self-efficacy variable (perceived behavioral control), and has been applied to under-standing health behavior (McEachan, Conner, Taylor, & Lawton, 2011). TPB has been the focus of numerous proposals to expand the list of variables that determine behavior, including the frequency of past behavior (Ouellette & Wood, 1998), automatic habits (Verplanken, 2006), and a variety of normative beliefs (Armitage & Conner, 2001). SCT also provides a compatible model of behavior governed by expected outcomes of behavior, self-efficacy, and self-regulatory norms, which has informed our own online safety research, as described below.

EVT and SCT also contributed to a stream of research in the MIS litera-ture, beginning with the Technology Acceptance Model (TAM, Davis, Bagozzi, & Warshaw, 1989) and its successors that were inspired by the Theory of Reasoned Action (Fishbein & Ajzen, 1975), another descendant of EVT and the predecessor of TPB. TAM has evolved into the Universal Theory of Adoption and Utilization of Technology, currently in its second iteration (UTAUT2; Venkatesh, Thong, & Xu, 2012). As with PMT, this research focused on variables specific to the behav-ioral domain in question, which was originally the acceptance of computer tech-nology in organizations, but that has subsequently been applied broadly across information technology and services and to consumer as well as organizational contexts. In the most recent formulation, outcome expectancies are encapsulated in the performance expectancy variable and the efficacy variable is called effort expectancy. Habits, social influence (parallel to perceived social norms in TPB and TRA), hedonic motivation (similar to benefits of maladaptive behavior in PMT), price value (related to cost of adaptive behavior), and facilitating conditions (e.g., availability of an effective response) are further predictors of behavioral intentions. UTAUT models also specify age, gender, and prior experience with a behavior as moderators of adoption intentions.

Age differences have been found in PMT research, although only when com-paring children to adolescents and adults (Rogers & Prentice-Dunn, 1997). Given the high prevalence of youth who are online (Pew Research Internet Project, 2014) and their views on privacy (Davis & James, 2013; Ito et al., 2008), the "digital natives" who grew up with the Internet are also of special interest in the online safety realm. Yet, older adults are the fastest growing segment of the online com-munity (Zickhur & Madden, 2012). The surging numbers of older adults going online constitute a new group of novice users who must learn to master online security threats. Retirement-age adults are more vulnerable to spam attacks, are a target of online scams, experience disproportionately high levels of identity theft, and take fewer defensive actions to protect themselves than younger users (Copes, Kerley, Huff, & Kane, 2010; Grimes, Hough, & Signorella, 2007; Senior Magazine Online, 2009). Unlike younger users who have access to online security resources through school or work, older adults are likely to be isolated home users and are less knowledgeable of online hazards (Grimes et al., 2007).

TAM-related theories have been applied in MIS research examining online safety in the context of risky online transactions (Chen, Wang, Herath, & Rao, 2011; Luarn & Lin, 2005; Pavlou, 2003; Yu, 2012). In this research, intention to adopt online banking services is the dependent variable and variables that reduce the risk of engaging in online banking are predictors. Others have directly examined online safety behaviors (Anderson & Agarwal, 2010; Herath, Chen, Wang, Banjara, Wilbur, & Rao, 2014; Johnston & Warkentin, 2010; Lai et al., 2012; Liang & Xue, 2010). Although the terminology varies between studies (e.g., safeguard effectiveness as opposed to response efficacy), all of the basic elements of threat and coping appraisals propounded in PMT are represented in this research. With the exception of Liang and Xue (2010), not all PMT variables are present in any one study.

Additional predictors of online safety behavior have been put forth in the MIS literature that might inform further development of PMT. Perceived security support (Liang & Xue, 2010) is a self-efficacy variable, and also an example of facilitating conditions in UTAUT2, referring to the availability of technical help to enact protections. Subjective norms (Anderson & Agarwal, 2010) are defined in terms of perceived moral (as opposed to technical) support for online safety, also found in TRA, TPB, and UTAUT2. Perceived citizen efficacy (Anderson & Agarwal, 2010) relates individual protective action to the overall safety of the Internet and is conceptually similar to the personal responsibility variable investigated in our own research (see below) and also to other protective (i.e., socially responsible) behavior uncovered in the PMT tradition (Wallerstein & Sanchez-Merki, 1994). The perceived credibility of the protections afforded by online partners (Luarn & Lin, 2005) and trust in online institutions (Pavlou, 2003) are relevant in high-risk transactions; for example, with online banks. These variables might be construed as further elements of coping appraisals within PMT.

Online safety research has provided further support for PMT. For example, security self-efficacy (coping self-efficacy in PMT), concern about security threats (threat susceptibility), and perceptions that individual Internet citizens protect the Internet (personal responsibility) were predictors of attitudes toward security related behavior (Anderson & Agarwal, 2010). Liang and Xue (2010) showed that perceived threat, safeguard cost (i.e., response cost), safeguard effectiveness (response efficacy), and (coping) self-efficacy predicted avoidance motivation (i.e., intentions to use the software), which in turn predicted avoidance behavior (i.e., the use of anti-virus software). Experimental interventions found in the online safety research (Anderson & Agarwal, 2010; Johnston & Warkentin, 2010; Marett, McNab, & Harris, 2011) further enrich a tradition that has relied extensively on cross-sectional studies. For example, Johnston and Warkentin (2010) modeled protection intentions among those who received a fear-inducing threat message about spyware and found that coping self-efficacy and response efficacy were both positive predictors of intentions to adopt spyware protection as was social influence. Also, both coping self-efficacy and response efficacy reduced perceived threat severity, suggesting that coping appraisals interact with threat perceptions.

Qualitative research examining "folk models" of Internet security further support and extend PMT concepts (Rader, Wash, & Brooks, 2012; Vaniea, Rader, & Wash, 2014; Wash, 2010; Wash & Rader, 2011). In this research, social support and social learning from the experiences of others encourage safe behavior, while the costs of implementing protections, such as software updates, are barriers. Threat appraisals are muddied by misunderstanding viruses and malware to be "buggy software" or just "generically bad," while failing to recognize their connection to harmful criminal activity. Perceptions of the sources of threats are a new direction for online safety research: Users credit threats from "burglar" hackers who steal identities and financial information, but not so much from "graffiti artists" showing off their hacking skills or from criminals who target "big fish" or work for criminal organizations because users don't see themselves as targets of the latter groups.

Other online safety research challenges well-established PMT findings. (Coping) self-efficacy in evaluating email messages had an inverse effect on intentions to adopt an email authentication service whereas response efficacy and threat levels were directly related to adoption (Herath et al., 2014). Liang and Xue (2010) found a negative interaction between safeguard effectiveness and perceived threat: Those who perceived a high level of threat were less motivated by the perceived effectiveness of safeguards than those with low levels of threat. In Anderson and Agarwal (2010), both threats and benefits were ineffective. Similar to health communication interventions stressing health tips, online safety interventions in the form of computer safety tips (e.g., "use firewalls") do not enhance and may even undermine the coping self-efficacy of the uninitiated (e.g., who do not know how to activate a firewall). Johnston and Warkentin (2010) examined coping self-efficacy as a mediator between threat perceptions and intentions. Liang and Xue (2010) modeled response efficacy, but not coping self-efficacy, as a moderator of threat perceptions.

Thus, similar sets of assumptions and theories inform online safety and health communication research. PMT is a useful starting point through which to examine the two streams of research, although they differ with respect to fine points of operational definitions as well as key findings regarding the strength and nature of relationships among key variables. Online safety research makes a contribution by offering new explanatory variables proposing new patterns of relationships among variables.

Testing PMT in the Online Safety Context: Empirical Research

In this section, we examine these issues in studies performed in recent years by our research team but which have not as yet been widely disseminated to communication scholars. A survey of 554 undergraduate students examined their perceptions of threat susceptibility, threat severity, coping self-efficacy, response efficacy, response rewards, safety habits, and personal responsibility norms in relation to online safety behavioral intentions (LaRose, Rifon, Liu, & Lee, 2005). Coping self-efficacy and response efficacy, but not other threat or coping appraisal variables, were significant positive predictors of online safety intentions in a model explaining 47% of the variance. Personal responsibility explained additional variance in online safety intentions over and above standard PMT variables: Those who considered online safety as their personal responsibility were more likely to perform online safety behavior, while those who think online safety was others' responsibility were less likely to carry out online safety behavior. The strength of safety protection habits was also a significant predictor of future protection intentions. Thus, personal responsibility and habit strength emerged as new coping resource variables while threat appraisals had relatively little impact.

A follow-up online experiment implemented a personal responsibility intervention with 206 undergraduate students (LaRose & Rifon, 2006; LaRose et al., 2008). In addition to supporting the main effect of personal responsibility manipulation on security intentions, they found a significant three-way interaction effect among personal responsibility (self vs. other), safety involvement (high vs. low), and self-efficacy (high vs. low) on security intentions. High personal responsibility attributions in persuasive messages advocating for online safety behaviors led participants to express greater intentions to perform online safety precaution behavior when controlling individual involvement and coping self-efficacy, in comparison with messages with low personal responsibility attributions. However, there was one exception: Those with low involvement in safety protection and low self-efficacy were less likely to take safety precautions when they were exposed to a high personal responsibility message than a low personal responsibility message. This research provided validation for the personal responsibility construct uncovered in the prior research and also exposed a possible boomerang effect in the online safety domain.

Vicarious experience was compared to verbal persuasion (after Bandura, 1994) to bolster coping self-efficacy in an online experiment involving 161 adult Internet users (LaRose, Rifon, & Wirth, 2007). Using a similar approach to LaRose et al. (2008), participants were randomly assigned to different conditions where they browsed a web page advocating for online safety. Elements of the web content were manipulated for personal responsibility attributions (self vs. others) and intervention style (persuasive vs. self-efficacy). Participants in the persuasion condition saw a webpage with several safety tips in text. Those in the vicarious experience self-efficacy condition saw step-by-step instructions by clicking on a "Show Me How" button in the webpage. Similar to the previous research, attributions of personal responsibility to the self were more effective at motivating protective behavior. There was a significant three-way interaction effect of personal responsibility (low vs. high), self-efficacy intervention style (vicarious experience vs. verbal persuasion), and prior knowledge (low vs. high) on online safety intentions. On one hand, for participants with low prior knowledge of online safety behaviors who were exposed to messages with high personal responsibility attributions, the vicarious experience intervention resulted in greater safety intentions than persuasion-focused interventions. Conversely, when stressing third-party responsibility to the low-knowledge group, the effect of self-efficacy backfired: The persuasion intervention was more effective than vicarious experience intervention. For people who had high prior knowledge of safety protections and were exposed to low personal responsibility messages (e.g., when they were told their ISPs were responsible for online safety), the vicarious experience intervention was more effective than persuasion intervention. However, no differences between vicarious experience and persuasion interventions were found for participants with high knowledge and who were exposed to high personal responsibility treatments. Thus, messages stressing personal responsibility and coping self-efficacy to promote online safety could be tailored to the level of involvement and prior knowledge of individual Internet users to increase their effectiveness.

A cross-sectional online survey of 998 Amazon Mechanical Turk workers living in the United States, who were broadly representative of the demographics of the national Internet population, explored factors that predict online safety behavior (Tsai et al., 2014). Nearly half of the variance (46.7%) in security intentions was explained by conventional PMT threat appraisals (i.e., threat severity, threat susceptibility, response costs), and an extended list of coping appraisals that included conventional PMT concepts (response efficacy and coping self-efficacy) and new additions from our own research and the MIS literature (i.e., prior experience with safety hazards, personal responsibility, safety habit strength, subjective norms, perceived security support, e-Trust, and perceived credibility of online service providers). Threat severity, but not threat susceptibility, influenced security intentions. Among coping appraisals, safety habit strength was the strongest positive predictor of security intentions, consistent with LaRose et al. (2005), followed by personal responsibility for online safety, e-Trust, subjective norms, and prior experience with safety hazards. Response efficacy was a positive predictor and response cost was a negative predictor of online safety (security) intentions. Unlike previous PMT research, coping self-efficacy was negatively associated with security intentions in the regression analysis, although there was a positive zero-order correlation, suggesting coping self-efficacy may have been suppressed by other variables in the regression model. This research extended previous research to an adult sample and provided evidence for a reconceptualization of coping appraisals to include external supports that may account for beliefs in one's personal abilities to overcome online hazards.

A second Mechanical Turk sample stratified by generational cohorts to include Millennials (born between 1977–1992), Older Boomers (born between 1946 and 1954) and Silent and GI generations (SGI; born before 1946) was recruited to examine differences in online safety perceptions in the context of online banking across the life span (Alhabash, Cotten, Tsai, Rifon, & LaRose, 2014). The generational cohorts significantly differed in online security perceptions and online banking attitudes and intentions. Millennials had higher levels of performance expectancy, effort expectancy, coping self-efficacy, response efficacy, perceived security support, trust, habit strength, and more favorable online banking attitudes and intentions but lower levels of threat susceptibility and response costs than older adults. While the regression model explained nearly half of the variance for the entire sample (without consideration of age cohort membership), the regression model for Boomers explained the greatest variance at 53%, followed by SGIs at 44%, and Millennials at 27%. Trust in the online banking institution was the most salient predictor of online banking adoption among the three cohorts. Performance expectancy was a positive predictor and response efficacy a negative predictor for the SGI and the Millennial groups. Similarly, coping self-efficacy was a negative predictor for the Boomers and the Millennials, while online banking attitudes and effort expectancy were significant positive determinants of online banking adoption specifically for the Boomer group.

An essential part of motivating Internet users to follow safe online practices is communicating accurate information in a way that the users can understand what to do, and that they will be motivated to take appropriate action. We carried out an analysis of nine websites dedicated to promoting online safety using basic constructs from PMT, health belief model (HBM), TPB, and SCT (Shillair, Boehmer, Cotten, LaRose, & Rifon, 2014). These web safety sites are sponsored by educational, volunteer, and governmental organizations. In general, the nine websites followed the key elements of the standard PMT model and introduced some of the extended list of factors that may influence adoption of online safety precautions: susceptibility to threats, severity of the threat, benefits of following safety precautions, self-efficacy of the user, response efficacy, general orientation or habit, social norm, and moral norm. However, the emphasis was clearly on threat appraisals that have a mixed record of success in the online safety domain, as well as elsewhere. With few exceptions, coping self-efficacy communication strategies involved simple exhortations to adopt safety tips or to avoid unsafe practices without further empowering users to execute those actions.

DISCUSSION

In this chapter, we provided an overview of a risk communication issue with an ever-growing complexity and relevance to understanding the ways in which communication can aid achieving good and healthy lives, whether for our physical selves or our online selves. The research we reviewed here points to an important and not-so-new issue related to online security and safety: People might understand that there is risk in using the Internet, they can sometimes assess their susceptibility to that risk, yet in many cases they fail to do what's necessary (or the bare minimum) to protect themselves while using the Internet and their computers.

Online safety research offers some extensions to PMT and related models of risk communication. It has investigated a wide array of resources other than response efficacy and coping self-efficacy that can influence individuals to adopt adaptive behavior. These include social support and observational learning from others, reliance on trusted institutions to provide protections, social norms that encourage safe behavior, and the formation of safety habits. New possibilities for the relationships between threat and coping appraisals have also been uncovered here, including the possibility that coping appraisals act as moderators rather than, or in addition to, mediators or direct precursors of adaptive behaviors. These issues would seem to have obvious parallels in other domains of risky behavior that deserve further investigation.

But also, some fundamental challenges to PMT are presented. After accounting for the full range of social and personal resources that individuals can summon to protect themselves, is coping self-efficacy still powerful? Or, do coping self-efficacy perceptions merely summarize one's ability to organize those resources into an effective response? Understanding more about the sources of coping ability might help to develop more effective interventions. And, how to explain the null and conflicting findings with respect to threat appraisals? By displacing the focus of the risk from the physical self to a virtual self or indeed to an external object are threats nullified? Or, does computer security represent such complex threats that denial and displacement occur even at moderate levels?

The issue of online safety is not one that is going to have a miracle solution in the near or far futures. Its complexity will continue to grow in parallel with the increasing sophistication and complexity of the Internet, as well as our increasing dependency on the Internet to socialize, entertain ourselves, bank, and shop. The stronger our habitual use of the Internet, the greater the risk that our understanding of the online threats are outweighed by the gratifications we get from using these platforms, the benefits of maladaptive behavior in PMT terms. While other fields, like computer science and human computer interaction, have been at the forefront of diagnosing the problem, we, as communication researchers, can dig deeper into understanding how to best communicate about online safety in a way that yields the desired changes in attitudes and behaviors.

Future research situated in the communication field should illuminate the understanding of online safety and security by doing what it does best: aggregating different fields to gain a comprehensive view of a phenomenon. Using models from psychology, sociology, computer science, and other fields, and utilizing interdisciplinary approaches, communication researchers can help with navigating this phenomenon at levels ranging from the micro to the macro levels.

The majority of PMT research applied to online safety has relied on retrospective, self-report measures of threat and coping appraisals. There is a role for media psychology and psychophysiology in understanding how users deal with online security threats to be able to understand how online safety messaging can use cognitive and emotional responses as predictors of future safety behaviors. A more in-depth inspection of messaging strategies will continue to be relevant in this regard. Another limitation is dependence on cross-sectional interviews, be they in the form of survey or qualitative interviews. There is a need for integrating cross-sectional measurement with longitudinal and behavioral data. This also extends to realizing the need to step outside the controlled laboratory when experimenting with intervention types, messages, and the like. More field experimentation is needed, yet it is of vital importance to balance such endeavors with responsible conduct of research. For example, observation of public behavior in relation to online safety (e.g., looking at individual's security update logs) can be beneficial and could be furthered by investigating effectiveness of certain interventions and message strategies. Future research should build on past research highlighting the importance of personal responsibility, coping self-efficacy, and institutional trust to explore how messaging strategies could lead individuals to adopt risky online behaviors, as well as take protective actions to guard against online threats. The link between cognitive and emotional responses to intervention strategies, on one hand, and actual behaviors could be an important contribution to this stream of research.

Implications for Online Safety Interventions

New approaches to online safety education are needed. Many of the online educational resources that average users are likely to encounter trade in safety "tips" and fear appeals that many do not have the required level of knowledge or confidence to enact and that confirm maladaptive forms of fear control, such as ignoring the problem or rationalizing away burdensome protective measures.

Age-appropriate targeting is one approach that is clearly called for. Although some members of older generational cohorts are tech savvy and cognizant of how to maintain online safety, the majority of older adult users do not have sufficient knowledge and skills to successfully maintain their online safety. Results from a large-scale randomized controlled trial illustrate that online safety, security, and privacy are primary concerns for older adults, concerns that can be addressed through focused training activities that are tailored to older adults (Cotten et al., 2013).

Although skills continue to increase as people gain more experience, technologies also continue to evolve, with an ever-increasing array of online safety threats. Understanding how different generational cohorts view online safety and security is one step in beginning to develop tailored interventions that will be effective for these different generational cohorts. However, across all generations, online safety education must be tailored to the entry-level skills and coping resources available to individuals. Interactive educational tools that assess entry levels, direct users to appropriate resources, and build competency through emulation and enactive mastery are both possible and necessary in the online environment. Online safety is clearly a moving target, with new threats constantly emerging that defy immediate technical solutions or even detection. Communicating accurate information about how to counter online threats is necessary but not sufficient. The ultimate goal of online safety education must be to imbue all users with a sense of vigilance and mastery that will motivate them to continually expand their own knowledge and effectiveness and to take advantage of the coping resources available to them as individuals.

Implications for Public Policy

Public policies might be enacted to encourage self-motivated vigilance and protective behavior, but little has been done to date. Online safety-related policies such as the Children's Online Privacy Protection Act (COPPA, 1998), the Children's Internet Protection Act (CIPA, 2000), and anti-"sexting" and anti-bullying laws now being enacted by many states focus on protecting children and adolescents by prohibiting access to inappropriate content. COPPA was revised in 2012 to request Internet content providers to acquire parents' approval before gathering data from users under age 13. These policies have indirect effects on the online safety of users, by reducing the risk that children will visit malware-laden web sites. However, many children under age 13 and even some parents submit a fake age in order to apply for an account (boyd, Hargittai, Schultz, & Palfrey, 2011).

Online safety policies with more general application have also focused on privacy protection and other kinds of informational threats to users, including SPAM (the CAN-SPAM Act), phishing (Anti-Phishing Act of 2005; Texas

Business and Commerce Code §325), identity theft (Identity Theft Penalty Enhancement Act of 2004; Identity Theft Enforcement and Restitution Act of 2008), and hacking (e.g., N.Y. Penal Law § 156.00; Cal. Penal Code § 502). The general thrust of these policies is to criminalize and punish purveyors of offensive or destructive content. Lacking are policy and public education initiatives targeting the behavior of individual users and standards for effective protections, such as exist in the public health and traffic safety domains. After all, if individuals are to believe themselves effective in coping with online threats, they need to be assured that there are effective means for responding to them and that others are likely to enact them as well. Also, public policies have not as yet addressed the need for effective responses to online safety threats on new platforms, such as mandating "kill switches" on smartphones.

The first step should be trials promoting effective online safety education practices. Given the current legislative gridlock at the Federal level, state-level initiatives are needed. For example, the State of Michigan organized a cyber safety initiative through the state attorney general's office to distribute grade-appropriate educational materials to participating school districts throughout the state. Once effective programs have been developed and tested, compulsory education should be extended. For example, Michigan has online courses to support licensing and certification in the areas of boating, hunter, and snowmobile safety as well as mandatory driver education. While the faintest suggestion of "drivers licenses" for Internet users would undoubtedly set off a storm of protest, it should be possible to require training and certification for the use of public computing facilities, including those in schools and libraries. Drawing another analogy from highway safety, access to online public services could be limited to those whose computers pass a safety inspection to assure that current versions of browser patches are installed, for example.

As we have shown in this chapter, much theoretical and empirical work has been conducted over the past three decades on this topic. However, much yet remains to be done theoretically, empirically, and policy-wise. We encourage other researchers and policy-makers to use the information provided in this chapter to expand efforts to promote effective online safety practices for all individuals and organizations.

NOTE

1. Acknowledgment: This research was supported by NSF grants #0430318 and #1318885 and by a grant from the Microsoft Foundation. The opinions expressed are those of the authors.

REFERENCES

Ajzen, I. (1991). The theory of planned behavior. *Organizational Behavior and Human Decision Processes, 50*, 179–211.

Alhabash, S., Cotten, S., Tsai, H. S., Rifon, N. J., & LaRose, R. (2014, August.) *Understanding generational differences in the relationship between online banking and online security.* Paper presented at the Association for Education in Journalism and Mass Communication, Communication Technology Divisions, Montreal, Canada.

Anderson, C. L., & Agarwal, R. (2010). Practicing safe computing: A multimedia empirical examination of home computer user security behavioral intentions. *MIS Quarterly, 34*(3), 613–643.

Anti-Phishing Act of 2005, CAL. BPC. CODE § 22948.2.

Armitage, C. J., & Conner, M. (2001). Efficacy of the theory of planned behaviour: A meta-analytic review. *British Journal of Social Psychology, 40*(4), 471–499.

Atkinson, J. W. (1964). *An introduction to motivation.* Princeton, NJ: Van Nostrand.

Bandura, A. (1977). Self-efficacy: Toward a unifying theory of behavioral change. *Psychological Review, 84*(2), 191–215.

Bandura, A. (1986). *Social foundations of thought and action: A social cognitive theory.* Englewood Cliffs, NJ: Prentice-Hall.

Bandura, A. (1994). *Self-efficacy.* In R. J. Corsini (Ed.), *Encyclopedia of psychology* (2nd ed., Vol. 3, pp. 368–369). New York: Wiley.

boyd, d., Hargittai, E., Schultz, J., & Palfrey, J. (2011, November 7). Why parents help their children lie to Facebook about age: Unintended consequences of the 'Children's Online Privacy Protection Act.' *First Monday, 16*(11). Retrieved from http://firstmonday.org/ojs/index.php/fm/article/view/3850

Cal. Penal Code § 502.

Campis, L. K., Prentice-Dunn, S., & Lyman, R. D. (1989). Coping appraisal and parents' intentions to inform their children about sexual abuse: A protection motivation theory analysis. *Journal of Social and Clinical Psychology, 8*(3), 304–316.

CAN-SPAM Act of 2003, Pub. L. 108–187, codified at 15 U.S.C. § 7706(f)(1).

Chen, R., Wang, J., Herath, T., & Rao, H. R. (2011). An investigation of email processing from a risky decision making perspective. *Decision Support Systems, 52*(1), 73–81.

Children's Internet Protection Act (CIPA) of 2000, Pub. L. 106–554. 20 U.S.C. §§ 6801, 6777, 9134 (2003), codified at 47 U.S.C. § 254.

Children's Online Privacy Protection Act (COPPA) of 1998, Pub. L. 105–277, 112 Stat. 2681, (October 21, 1998), codified at 42 U.S.C. §§ 6501–6507.

Copes, H., Kerley, K. R., Huff, R., & Kane, J. (2010). Differentiating identity theft: An exploratory study of victims using a national victimization survey. *Journal of Criminal Justice, 38*(5), 1045–1052.

Davinson, N., & Sillence, E. (2010). It won't happen to me: Promoting secure behaviour among internet users. *Computers in Human Behavior, 26*(6), 1739–1747.

Davis, F. D., Bagozzi, R. P., & Warshaw, P. R. (1989). User acceptance of computer technology: A comparison of two theoretical models. *Management Science, 35*(8), 982–1003.

Davis, K., & James, C. (2013). Tweens' conceptions of privacy online: Implications for educators. *Learning, Media and Technology, 38*(1), 4–25.

Fishbein, M., & Ajzen, I. (1975). *Belief, attitude, intention and behavior: An introduction to theory and research*. Reading, MA: Addison-Wesley.

Grimes, G. A., Hough, M. G., & Signorella, M. L. (2007). Email end users and spam: Relations of gender and age group to attitudes and actions. *Computers in Human Behavior, 23*(1), 318–332.

Herath, T., Chen, R., Wang, J., Banjara, K., Wilbur, J., & Rao, H. R. (2014). Security services as coping mechanisms: An investigation into user intention to adopt an email authentication service. *Information Systems Journal, 24*(1), 61–84. doi: 10.1111/j.1365-2575.2012.00420.x

Identity Theft Enforcement and Restitution Act of 2008, Pub. L. No. 108–275, Tit. II, 122 Stat. 356 (Sept. 26, 2008), codified at 18 U.S.C. §1030.

Identity Theft Penalty Enhancement Act of 2004, Pub. L. No. 108–275, 118 Stat. 831 (July 15, 2004), codified at 18 U.S.C. §§1028, 1028A.

Ifinedo, P. (2012). Understanding information systems security policy compliance: An integration of the theory of planned behavior and the protection motivation theory. *Computers & Security, 31*(1), 83–95.

Ito, M., Horst, H., Bittanti, M., boyd, d., Herr-Stephenson, B., Lange, P. G.,… Robinson, L. (2008). *Living and learning with new media: Summary of findings from the digital youth project*. Chicago, IL: John D. and Catherine T. MacArthur Foundation.

Johnston, A. C., & Warkentin, M. (2010). Fear appeals and information security behaviors: An empirical study. *MIS Quarterly, 34*(3), 549–566.

Lai, F., Li, D., & Hsieh, C. T. (2012). Fighting identity theft: The coping perspective. *Decision Support Systems, 52*(2), 353–363.

LaRose, R., & Rifon, N. (2006, June). *Changing online safety behavior: Experiments with online security and privacy*. Paper presented to the International Communication Association, Dresden, Germany.

LaRose, R., Rifon, N., & Enbody, R. (2008). Promoting personal responsibility for online safety. *Communications of the ACM, 51*(3), 71–76.

LaRose, R., Rifon, N., Liu, S., & Lee, D. (2005, May). *Understanding online safety behavior: A multivariate model*. Paper presented to the International Communication Association, Communication and Technology Division, New York, NY.

LaRose, R., Rifon, N., & Wirth, C. (2007, May). *Online safety begins with you and me: Getting Internet users to protect themselves*. Paper presented to the International Communication Association, San Francisco, CA.

Lee, D. W., LaRose, R., & Rifon, N. (2008). Keeping our network safe: A model of online protection behaviour. *Behavior & Information Technology, 27*(5), 445–454.

Liang, H., & Xue, Y. (2010). Understanding security behaviors in personal computer usage: A threat avoidance perspective. *Journal of the Association for Information Systems, 11*(7), 394–413.

Luarn, P., & Lin, H.-H. (2005). Toward an understanding of the behavioral intention to use mobile banking. *Computers in Human Behavior, 21*(6), 873–891. doi: 10.1016/j.chb.2004.03.003

Marett, K., McNab, A. L., & Harris, R. B. (2011). Social networking websites and posting personal information: An evaluation of protection motivation theory. *AIS Transactions on Human-Computer Interaction, 3*(3), 170–188.

Martin, N., & Rice, J. (2013). Spearing high Net wealth individuals: The case of online fraud and mature age Internet users. *International Journal of Information Security and Privacy* (IJISP), *7*(1), 1–15.

McEachan, R. R. C., Conner, M., Taylor, N. J., & Lawton, R. J. (2011). Prospective prediction of health-related behaviours with the theory of planned behaviour: A meta-analysis. *Health Psychology Review, 5*(2), 97–144.

National Cyber Security Alliance, McAfee, & JZ Analytics. (2012). *2012 NCSA / McAfee Online Safety Survey.* Retrieved from http://staysafeonline.org/stay-safe-online/resources/

N.Y. Penal Law § 156.00.

Ouellette, J. A., & Wood, W. (1998). Habit and intention in everyday life: The multiple processes by which past behavior predicts future behavior. *Psychological Bulletin, 124*(1), 54–74.

Pavlou, P. A. (2003). Consumer acceptance of electronic commerce: Integrating trust and risk with the technology acceptance model. *International Journal of Electronic Commerce, 7*(3), 101–134.

Pew Research Internet Project. (2014). *Internet user demographics, teen Internet access demographics.* Retrieved from http://www.pewinternet.org/data-trend/teens/internet-user-demographics/

Rader, E., Wash, R., & Brooks, B. (2012, July). Stories as informal lessons about security. Proceedings from *The Eighth Symposium on Usable Privacy and Security (SOUPS).* New York: ACM.

Reeves, B., & Nass, C. (1996). *The media equation: How people treat computers, television, and new media like real people and places.* Stanford, CA: CSLI.

Rogers, R. W. (1975). A protection motivation theory of fear appeals and attitude change. *The Journal of Psychology, 91*(1), 93–114.

Rogers, R. W., & Prentice-Dunn, S. (1997). Protection motivation theory. In D. S. Gochman, (Ed.), *Handbook of health behavior research 1: Personal and social determinants* (pp. 113–132). New York: Plenum Press.

Seidman, D. (2012). Why Johnny can't patch and what we can do about it. Presentation to *Bluehat 2013.* Redmond, Washington. Retrieved from http://channel9.msdn.com/Events/Blue-Hat-Security-Briefings/BlueHat-Security-Briefings-Fall-2012-Sessions/BH1209

Senior Magazine Online. (2009). Internet scams: Fraud on the Internet. *Senior Magazine Online.* Retrieved from http://www.seniormag.com/legal/internet-scams.htm

Shillair, R., Boehmer, J., Cotten, S. R., LaRose, R., & Rifon, N. (2014, May). *Communicating online safety: A critical content analysis of websites providing online safety information to consumers over time.* Paper presented at the International Communication Association's 64th Annual Conference, Seattle, WA.

Texas Business and Commerce Code §325, Texas Statutes § 325.

Tsai, H. S., Jiang, M., Alhabash, S., LaRose, R., Rifon, N., & Cotten, S. R. (2014, May). *Understanding online safety behavior in the online banking context.* Paper presented at the International Communication Association's 64th Annual Conference, Seattle, WA.

Vaniea, K., Rader, E., & Wash, R. (2014, May). *Mental models of software updates.* Paper presented at the International Communication Association's 64th Annual Conference, Seattle, WA.

Venkatesh, V., Thong, J. Y., & Xu, X. (2012). Consumer acceptance and use of information technology: Extending the unified theory of acceptance and use of technology. *MIS Quarterly, 36*(1), 157–178.

Verplanken, B. (2006). Beyond frequency: Habit as mental construct. *British Journal of Social Psychology, 45*, 639–656.

Wash, R. (2010, July). Folk models of home computer security. Proceedings from *The Sixth Symposium on Usable Privacy and Security (SOUPS).* New York: ACM.

Wash, R., & Rader, E. (2011, September). Influencing mental models of security: A research agenda. Proceedings from *The 2011 Workshop on New Security Paradigms Workshop* (pp. 57–66). New York: ACM.

Wallerstein, N., & Sanchez-Merki, V. (1994). Freirean praxis in health education: Research results from an adolescent prevention program. *Health Education Research, 9*(1), 105–118.

Yu, C. S. (2012). Factors affecting individuals to adopt mobile banking: Empirical evidence from the UTAUT model. *Journal of Electronic Commerce Research, 13*(2): 104–121.

Zickhur, K., & Madden, M. (2012). Older adults and Internet use. *Pew Research Internet Project.* Retrieved from http://pewinternet.org/Reports/2012/Older-adults-and-internet-use.aspx

PART III

Challenges, Opportunities, and Transformation

CHAPTER ELEVEN

Communicative Figurations of the Good Life

Ambivalences of the Mediatization of Homelessness and Transnational Migrant Families

ANDREAS HEPP, UNIVERSITY OF BREMEN, GERMANY

PETER LUNT, UNIVERSITY OF LEICESTER, UK

MAREN HARTMANN, UNIVERSITY OF THE ARTS BERLIN, GERMANY

The concept of "mediatization" has been the focus of considerable debate and reflection for scholars in media and communication seeking to understand an increasingly media-related world (Couldry & Hepp, 2013; Hjarvard, 2013; Lundby, 2014). In theoretical work, mediatization is defined as transformations in media and communications that relate to social and cultural change as a societal meta-process akin to individualization, urbanization, and rationalization (Hepp, 2013a; Krotz, 2009). These reflections are increasingly complemented by empirical studies investigating transformations in institutions as well as social and cultural practices on different scales over varying historical periods. This includes the *longue durée* of human history, the consequences of media for modernity, and the more recent emergence of a mediated network society (Jensen, 2013; Livingstone, 2009; Livingstone & Lunt, 2014).

Much of this empirical research has examined the potential for media-influenced transformations in specific domains of life often focusing on particular media. In this chapter, in contrast, we emphasize the importance of understanding mediatization in the context of complex media environments and argue that a "communicative figurations" approach (Hepp, 2013a, pp. 92–97), based on Elias' process sociology, is a potentially useful framework to capture this understanding of mediatization. We use two examples to illustrate this point: the ontological insecurity of homelessness and the use of media by migrant mothers in transnational families. Both cases, although critical of hyperbole about digital media, examine the ethical potential of new media in connecting those disconnected through homelessness and enabling parenting at a distance. An analysis of these situations inevitably involves a variety of media rather than the operation of a particular technology. These two cases focus on the role of social media for everyday life in the context of contemporary late modern societies presenting a challenge to mediatization theory that focuses on processes of historical transformation. Livingstone and Lunt (2014) argue that "mediatization" refers to either long-term historical cultural change or to the role of media in modernity over recent centuries. In addition, the concept appears ideally suited to explain the contemporary media-related transformations in late modern societies against the backdrop of former forms of societies.

The challenge we confront in this article, therefore, is to relate empirical research on contemporary forms of life in the new media landscape to mediatization. In so doing our aim is to discuss whether mediatization is amenable to a "history of the present." Our main argument is that a process sociology approach—the aforementioned figurational perspective—is a highly helpful starting point to apply ideas about the mediatization of culture to contemporary forms of life. This leads us to consider whether an understanding of media ethics in both cases (insecurity of homelessness and media uses of migrant mothers) might enable us to link mediatization theory to normative questions of "the good life."

PROCESS SOCIOLOGY AND COMMUNICATIVE FIGURATIONS: THREE THINGS WE CAN LEARN FROM ELIAS

Sociologist Norbert Elias produced ground-breaking work through his book *The Civilizing Process* (Elias, 2000/1939), whereby he developed a sociology of "social processes" (Elias, 1978, p. 17) aimed at distinguishing structural transformation from the inherent dynamic changes in modern society. Three aspects of his work are particularly relevant to understanding transformation in present media and society: (1) the necessity of *linking* the individual with the social, (2) the *process* perspective, and (3) the distinction between *transformation* and social change.

First, the necessity of linking the individual with the social is fundamental to Elias' thinking. He criticizes the then (and still now) two dominant perspectives on the relationship between individuals and society: Either society is understood as the aggregated outcome of individual actions, or as a more or less autonomous system in which no account is taken of individual action unless the individual is conceived of as a similarly autonomous system (Luhmann, 2012). Elias (2000) argues for an approach that examines individuals in their complex and dynamic interrelationships with society, which is more than aggregated individual actions, and cannot be reduced to individual plans and purposes. Society also provides the context for action and a variety of constraints on what it means to be an individual and is the result of both individual and social transformation.

Second, Elias focuses sociological theory on processes rather than structures. A starting point was his criticism of key assumptions of sociological theories, "even the concept of social change is often used as if it referred to a fixed state—one drifts, so to speak, from seeing the state of rest as normal to seeing motion as a special case" (Elias, 1978, p. 115). Media and communication research is similarly obsessed with "change," having a tendency to focus on the latest media developments and innovations. However, Elias' request for a process perspective is more fundamental, proposing an understanding of both individual existence and the existence of society as ongoing processes of originating and constructing. From such a point of view, inertia is not a "fixed state," but part of an ongoing process of rearticulating a social environment constituted by the interplay of psychological and sociological transformations. The important theoretical and methodological implications of these insights are that we should focus on the processes of constructing the psychological and the social, instead of taking these as something given. Here, Elias' arguments are reflected in more recent developments within media and communication research, which ask for a "practice" (Couldry, 2004) and "communicative constructivist" (Knoblauch, 2013) analysis of media and communication.

Third, connected to this process perspective, and particularly pertinent to the contemporary context, is Elias' analysis of the distinction between transformation and social change. Change is the norm in modern society, and everything is in a constant state of flux in contrast to more fundamental social transformation that arises from dynamic social change. Examples in sociological theory are processes of individualization, urbanization, and rationalization. History suggests that to live in times of "acceleration" (Rosa, 2013) is not exclusive to late modernity and to mobile phones and other recent technologies, but a general impression that has also accompanied the emergence of "slower," linear media (cf. Tomlinson, 2007). Indeed, the feeling of living in times characterized by dynamic change has a long history. Simmel (1971/1903), for example, wrote elegantly of the emerging metropolis at the end of the 19th century in Europe as an environment in which dynamic social interaction created a new structure of consciousness and apparently a chaotic society. Therefore, it is no easy task for critical analysis to work out the more fundamental transformations that lie beneath the surface impressions of change that accompany modernity.

A Communicative Figurations Approach

Elias identifies two problems for the sociology of modernity: the relative autonomy but co-dependence of individuals and society, and the distinction between social change and structural transformation. His solution was to argue that structural transformation could be explained in terms of the shifting relation between individuals and society over time; this, he called figuration. Figurations are "networks of individuals" (Elias, 1978, p. 15) which constitute a larger social entity through processes of human practice and interaction. It is a "simple conceptual tool" (Elias, 1978, p. 30) used to understand sociocultural phenomena in terms of "models of processes of interweaving" (Elias, 1978, p. 130). And transformation can be understood as the change from one figuration to another.

We suggest that the notion of figuration can be used to explain changes in media and communication as *communicative figurations* (Hepp, 2013b; Hepp & Hasebrink, 2014). These are patterns of interweaving processes across various media platforms that combine a thematic framing that orients communicative action (Hepp, 2013b), and that can be identified at different scales. For example, a micro-level example is the way that migrant families are separated in space but connected through multimodality such as letters, the (mobile) phone, and social web through communication channels that keep family relationships alive. A meso-level example of communicative figurations of social organizations such as databanks may include the Internet as well as printed flyers and other public relations (PR) media intertwined to create its social order. A macro-level example of communicative figurations of public spheres might consist of a wide range of traditional mass media and emerging online outlets (Hasebrink & Hölig, 2014).

A communicative figuration has four features (Hepp & Hasebrink, 2014). First, we are dealing with *forms of communication*, the patterns of communicative practices. Second, each communicative figuration refers to a specific *media ensemble*, or a set of platforms and services that are involved in the communicative practices that constitute the figuration. Third, each communicative figuration is characterized by a *constellation of actors* that can be regarded as its structural basis, a network of individuals being interrelated with each other. Fourth, each communicative figuration has a *framing* (Goffman, 1974) that serves to guide concrete actions and practices such as those in migrant families, social organizations, and public spheres.

Grasped in this way, communicative figurations have constructive capacity. They are the means by which we communicatively produce *rules* (both resources and constraints) and construct our social identities and sense of *belonging* to various communities. Meanwhile, they also *segment* individuals and groups by excluding them from spheres of communicatively constructed social reality. Moreover, communicative figurations are marked by *power*. They work as forces of empowerment and disempowerment through the way that they give voice to some individuals or groups and exclude others. However, communicative figurations are not "given," but articulated in an ongoing process. Within this constitutive process communicative figurations change and may become part of deeper processes of transformation.

Media Ethnics and "The Good Life"

How are these reflections on communicative figurations connected to questions of the relationship between media and "the good life"? First, figuration reminds us of the complexity of the media environment that an individual is embedded in, and that the change of a particular medium is far less important than the potential transformation of the communicative figurations they are engaged in. Furthermore, this perspective allows us to identify at least three types of transformation in media that are potentially part of communicative figuration. First is an increasing *multi-optionality* arising from the availability of an increasing variety of new media technologies that interweave to constitute the technical infrastructure of communicative figurations. In this, individuals are "plural actors" (Lahire, 2011/2001) who are engaged with a variety of "particular inclusions" (Burzan, Lökenhoff, Schimank, & Schöneck, 2008) in various communicative figurations. Second, there is an increasing *mediacy* (Schütz, 1967) in which mediated social practices are distributed in space and time. For example, increasingly global media mean that communicative figurations are often translocal, extending across various localities. Third, *asynchronicity* refers to the continuing importance of inequalities in access and use of communicative figurations.

To what extent would these communicative figurations and transformations support the needs of this person in the pursuit of a good life? Couldry (2012, 2013) argues that with the increasing mediatization and related media saturation of our contemporary lives, this means that virtuous social practice raises questions involving us all, and not just media professionals. Therefore, the ethical question we all face is, "How should we act in relation to media, so that we contribute to lives that, both individually and together, we would value on all scales, up to and including the global?" (Couldry, 2012, p. 189). The ethical implications of communicative figurations therefore raise the question: "How should we act in relation to certain communicative figurations, so that we contribute to lives that, both individually and together, we would value on all scales?"

One way to approach this question is to consider the various needs a person has in such a context (Couldry & Hepp, 2012). Adopting a conception of human needs based on general human "capabilities" (cf. Sen, 1992, 1999), we see needs as socially constructed and shaped by the common pressures of material and historical conditions. Couldry (2012, pp. 163–179) offers the example of seven fundamental needs based on these assumptions: "economic needs" (related to economic security), "ethnic needs" (the togetherness in ethnic groups), "political needs" (political inclusion and participation), "recognition needs" (reflecting social "acceptance" within various contexts), "belief needs" (concerning the field of religion), "social needs" (those of social connection), and "leisure needs" (recreation). While these needs intersect and might be extended based on further empirical research, these examples offer a point of departure to reflect on the way that particular communicative figurations might enable or constrain the satisfaction of such needs by affording the relevant capability.

EMPIRICAL ANALYSES OF COMMUNICATIVE FIGURATIONS

Ontological Security and the Communicative Figurations of Homeless People

"Economic needs" are often at the forefront of discussions of homeless people as they shape their everyday life experiences. However, we would argue that their "social needs," "recognition needs," and "political needs" are also equally important given their close link to communication (e.g., being spoken about, being recognized). As quoted in Roberson and Nardi (2010), Jackie, a 61-year-old homeless woman in Los Angeles, said, "So you need to have a cell phone. Most people can't afford it. People go out and pick up cans just so they can have something to eat, but these other things are necessities too" (p. 447).

It is not surprising, then, that in 2000 a politician in Berlin asked for mobile phones to be given to the homeless, and this was not greeted with much enthusiasm because the framing of needs in this case were primarily as "economic needs" (Spiegel, 2000). We argue that a focus on communicative figurations helps to broaden our understanding of needs as capabilities and to recognize that the homeless tend to be excluded from parts of the communicative construction of reality: They are *segmented* or excluded from communicative opportunities that might enable them to express their capabilities. Digital media use can potentially alleviate this problem and contribute to demonstrate broader understanding of the ethical potential of media use in contexts of social exclusion based on an analysis of the communicative figurations at play (e.g., resources and constraints) for a homeless person living in a media-saturated society.

This argument intersects with research on the perspective of homelessness and the experience of homelessness as ontological (in-)security (Hartmann, 2014). What is at stake here is the conceptual link between "home" and "security." Although homeless people by definition lack home and shelter, they also lack the feeling of belonging, or *ontological security*. According to Laing (1960), ontological security provides people with a sense of "presence in the world as a real, alive, whole, and, in a temporal sense, a continuous person," with a clear sense of individual and social identity and perceptions of reality. Ontologically insecurity, equally, can be defined as when a person has no "parents, home, wife, child, commitment, or appetite" (Laing, 1960, p. 39).

There is an interesting tension here in relation to Elias' distinction between change and transformation as in which "change" is the norm, and the expectation is that the next day will be different from the day before. In this context, ontological security is not about stability, but about the degree, complexity, or manageability of change, and the recognition of the potential of transformation in response to change. For example, Winnicott (1971) introduced the idea of the transitional object, which helps children to develop the idea of "not-me" and thereby to eventually be able to separate themselves from their parents (the feeling of ontological security originally provided by the mother is temporarily substituted by the object). The transitional object is important in terms of the development of the child's identity and thereby also of security. The term "transitional object" also implies that Winnicott did not think of ontological security as something fixed but rather as a process. This emphasis on security, then, is not necessarily a contradiction of the understanding of change as the norm, even if the apparent aim of everyday life puts security at the forefront. At the same time, there is potentially too much change ahead for a homeless person so that constant flux becomes a threat. Therefore, routines are also developed to keep at least the minimal level of reliability and continuity. One important aspect about ontological security is the need for a sense of place or home, because "having a house is viewed as a normative base from which to achieve ontological security and stability [...] because it is a place where tensions that build up from constant surveillance in other settings can be relieved" (Brueckner, Green, & Saggers, 2011, p. 3). Security, according to this interpretation, is about providing a place that acts as a secure foundation from which we can enter into and act on the world and be able to retreat from the world when necessary.

In relation to the new media environment, the question we ask is whether communicative figurations in which homeless people can construct their day-to-day routines and create connections might develop a sense of control and identity that is not in any sense a substitute for the security of place and home. The suggestion here is that digital media can partially offer a process for the development of a sense of security, a feeling of independence, and a context for self-development (Hartmann, 2014; Tomas & Dittmar, 1995). For example, studies have suggested that at least the younger homeless (especially in the U.S.) tend to use social media fairly extensively, including frequency of access, content consumption, and social networking (Rice, Monro, Barman-Adhikari, & Young, 2010). The differences lie more in the points of access and the need to invest (effort mostly) to get this access (Pollio, Batey, Bender, Ferguson, & Thompson, 2013, p. 174). As Woelfer and Hendry (2012) suggest, in addition to organizing their everyday life on the street, the exploration of identity and the cultivation and exploitation of social ties are at the forefront of the home inasmuch as "the social networks of homeless young people can be exploited for opportunity but, even more, for human well-being" (p. 7). This supports the idea that digital media, and especially social media, can at least in principle "help the individual and the collectivity to define and sustain their own ontological security wherever they happen to be" (Silverstone, 2006, p. 233), including the homeless. Social media can serve as the "transitional object" for at least some of the homeless, building up communicative figurations that help to construct the ontological security and potentially building a good life.

This argument also picks up the focus on exclusion from media life as communicative figurations outlined above, and the importance of a minimal ethics of communicative commitments through engagement in networks of social interaction. Indeed, one of the problems for many homeless people is their exclusion from many communicative figurations, which are taken for granted by others. All forms of life are partly constituted through communicative figurations, and homelessness is no exception. However, in the case of the homeless, most of these figurations are not mediated by the homeless themselves. The constructions of the sociocultural reality and symbolic meaning of the homeless are therefore limited (albeit differently so in different contexts). Communicative figurations of the homeless tend to be mostly exclusive. They work on an "us vs. them"-structure, or a differentiation between the established and the outsiders (Elias & Scotson, 1994). Communicative figurations play a part in the creation of such differences through segmentation. They can, however, potentially create conditions for inclusion through communication.

How does this conception of digital and mobile media provide the potential for ethical commitment and realization of the social self? One way is to provide opportunities for all to play an active, constitutive role in communicative figurations. We would probably need to become even more normative if we wanted to live up to the notion of an extended media ethics in this context, especially in relation to "social needs" and "recognition needs," for example, through a general public commitment to include the homeless more actively. In one of the few attempts to apply virtue ethics in the context of homelessness, Burkum (1999) argues, "First, homelessness is not just the condition of lacking a home in the sense of a 'roof over one's head.' It is the situation of one who does not participate in the 'sphere of membership' [...] It is the condition of not being acknowledged as belonging to society. [...] we act as if they do not even exist" (p. 79). Burkum goes on to argue that membership in the community is the basis for human existence, and non-membership, as in the homeless, violates ethics (see also Silverstone, 2006). Their inclusion, on the other hand, would be the virtuous thing to do and thereby provides an appropriate ethical stance. The community that Burkum speaks about, however, needs an ethically enabling environment in relation to "social" and "recognition needs."

To summarize, digital media may enable the homeless to overcome some of the risks to existence that they face. A potentially easier step of inclusion might work through to the offer of (somewhat) surveillance-free digital media access accompanied with training. If the homeless are not visible (as they tend to be overlooked), but are at the same time under constant surveillance and therefore nowhere at home, this could be an initial "hide-away" as well as a possibility to develop everyday routines and to build an additional identity. This could be seen as one initial gesture to extend membership in our existing communities to the homeless in Burkum's sense.

One example is the recent online protests against the architectural exclusion of the homeless through the erection of bumps in floor areas where the homeless sleep, known as the "anti-homeless-studs" in the public debate in UK. The widespread argument was that this kind of active exclusion was not acceptable, especially since the devices looked similar to those intended to deter pigeons. The original tweet that caused the outrage included a picture of the studs, and a statement saying "Anti homeless floor studs. So much for community spirit": (Ethical Pioneer, 2014). This reinforces the idea of community ideally aiming at something more communal. The particular user who introduced the tweet (he calls himself "Ethical Pioneer") clearly displays signs of being a virtuous character. In addition, because individual acts are never enough, the interplay of such individual virtuous acts together with communities of action and communicative figurations may begin to address the social problem of exclusion of the homelessness.

This example is less an example of the transformation of communicative figurations, but more of how critical changes in the life course of a person (becoming homeless) are related to radical changes of the communicative figurations the person is involved in. In this case, the media are not the driving forces of change, but they can help to make radical changes to insecure life situations more manageable for the people involved.

Social Relationships and the Communicative Figurations of Transnational Families

In a number of studies of Filipino families separated by migration, Madianou and Miller (2012) examined how mobile and digital media are used to sustain family relationships and to create content that represents the views and experiences of parenting at a distance. Madianou has also explored the broader context of migration as a social phenomenon with a variety of stakeholders including migrant families, governments, non-governmental organizations (NGOs) and telecom providers (see Madianou and Miller, 2012).

The question about migrant and transnational families that many scholars, including Madianou, are now asking is whether the combination of mobile phones, broadband connections, digital cameras, and social media help to artic-ulate communicative figurations that extend these transformations further than linear and traditional mass media managed to. In this sense, Madianou's study is an examination of contemporary forms of the mediation of family life that links to mediatization in two ways: in the sense that we can interpret the new communica-tive figurations of the family as being media-saturated social relationships that are, in significant ways, transformed by the extension of the capacity to communicate over space and time, and that the choices and actions of families take place in a broader social and institutional context involving government policies, NGOs, and providers of digital communication technologies.

The Philippines are a particularly appropriate context for such a study as the country has one of the highest per capita proportions of people working abroad, and this disproportionately affects women of childrearing age, creating a wide-spread phenomenon of "left-behind" children. In addition, the Philippines, along with other South Asian countries, have been at the forefront of developments in mobile technologies (Madianou & Miller, 2012).

Madianou's work initially came out of the emerging field of media and digital anthropology (Horst & Miller, 2012), and she emphasizes that the social change related to the appropriation of new media of migrant families is likely to be an example of cumulative cultural change rather than an example of mediatization as the influence of "media logics" on social institutions (Livingstone & Lunt, 2014). This works well for the dimension of her project focused on the mediation of family relations; however, her second focus on changing government policy, NGO activity, and the representation of migrants may be better thought of as an example of the institutional approach to mediatization (Hjarvard, 2013), and a particu-larly interesting one in which the aggregate factors (numbers of migrants) of a cumulative cultural change (distant family relationships) results in a shift in the orientation of social policy on migration. Here "recognition needs" have a high importance.

The family relations side of Madianou's study applies the traditions of eth-nographic research to the emerging field of media and digital anthropology. Madianou and Miller's (2012) work, then, provides an empirical demonstration of the adoption of diverse digital media technologies that contextualizes the fea-tures of the communicative figurations identified at the beginning of this article (forms of communication, a media ensemble, a constellation of actors, and a fram-ing of "family"). This communicative figuration furthermore demonstrates how we construct social relationships as family or motherhood as an ethic of care that addresses our "social needs." This is similar to suggestions made by Livingstone and Lunt (2014) that one way of linking the insights of mediatization theory to empirical studies of the media may be a collaborative venture with different aca-demic theories and studies of the different aspects of culture and society that are implicated in mediatization. A key reason for this is that the mapping of mediatization onto complex social phenomena requires a grounding that is sensitive to the complexity and contrariness of processes of change—as Madianou (2014) puts it, "Although the title of [this] chapter is 'the mediatization of migration' it is evident that migration is too complex and diverse a phenomenon for a single type of social change to occur" (p. 325).

Madianou (2014) outlines the development of the emerging field of media and migration as an interdisciplinary field of study. Traditionally there has been little research on media in migration research. This demonstrates, in a sense, one of the core features of how the contemporary developments of digital media and convergence culture have relatively penetrated in the societies so that the question of the role of media in migration can no longer be ignored. The focus of transnational media studies has traditionally been on production, textual, and audience studies of migration representations. Madianou sees most value in previous audience studies in which the complexities of identities in migration have been an interesting development (Georgiou, 2006; Siapera, 2005; Sreberny, 2005). Notwithstanding the difficulties of interpreting a complex, dynamic social and cultural phenomenon, such as migration in terms of identity, the advent of new media opens up a new research agenda focused on the practices of sustaining transnational relationships while preparing for migration, in enhancing existing separate relationships, and sustaining family life at a distance. This new research agenda extended to begin to examine how migration itself was being potentially changed through the adoption of new media communication technologies by separated families.

Madianou's starting point for this new research agenda is, within an ethnographic approach, to examine the way the adoption and use of new media might become part of the context within which participants establish social constructions of reality. A number of ecological metaphors have been proposed as potentially fruitful ways of understanding the cultural consequences of mediation, and Madianou and Miller (2013) offer a new conceptualization of media ecology as "polymedia." In essence, we are witnessing a transformation of the communicative figurations by which transnational migrant families (here, especially mothers and their children) develop their social relationships through their "social needs." This is a move from communicative figurations based on a small number of media partly with a time delay of communication (letters, audio cassettes) to communicative figurations of polymedia. The media ensemble of these figurations includes the mobile phone, Internet, telephony, etc. As a consequence, there is a shift "from a focus on the qualities of each particular medium as a discrete technology, to an understanding of new media as an environment of affordances" (Madianou & Miller, 2013, p. 170). Madianou adopts a focus on polymedia in which media are defined in relation to other media technologies in the context of the thesis of the social shaping of technology (Wajcman, 2002), in which technology and social relations are mutually constituted. This gives the ethnographic work a particular focus on the way in which people engage a variety of different combinations of media technologies, thereby transcending the affordances of discrete media (Hutchby, 2001) as people mix and match different combinations of mobile, digital, and social media to sustain their relationships at a distance.

Madianou (2014) interviewed mothers in the UK and young adult children in the Philippines in addition to a variety of policy and community stakeholders on migration, such as government and NGO representatives. A critical finding from the interviews with institutional stakeholders is optimism concerning the potential value of new media technologies for families separated by migration, by potentially sustaining family relationships, overcoming social problems related to separation, creating a market for mobile services, and enabling further migration (Madianou & Miller, 2012). This optimism was, to some degree, shared by the migrant mothers in offering what they regarded as intensive mothering at a distance achieved through more frequent points of contact and the capacity to "see" children (via, for example, Skype), and greater knowledge of the minutiae of their children's lives, addressing the "social needs" of the migrant mothers. At the same time, increased contact and knowledge of the circumstances back home could also lead to conflicts, such as the lack of ability to deal with problems back home, and a visual co-presence was often accompanied by a reminder of physical distance, creating emotional challenges (Madianou, 2014).

For the children and other family members left behind, a number of factors influenced the practices and quality of relationships sustained partly through new media, such as the age of the child at separation and the quality of the existing relationship between mother and child. Younger children and those who had existing relationship problems found less value in new media. There were clear distinctions between families that adopted and found value in new media in these circumstances, and those that did not, and between the accounts by transnational families and the stakeholders' optimism about new media.

Madianou links these findings to mediatization by suggesting that a number of unintended consequences of these patterns of adoption can be seen to affect migration. The consequences of new media are best understood at the level of the communicative figurations of everyday life, of the taken-for-granted interaction rituals of mediatized life. In addition, decisions to migrate for work and to extend the stay in the UK are being influenced by the potential of new media as reflected in policy discourse rather than by the lived realities of mediated family life at a distance.

Madianou and Miller's (2012) work therefore illustrates both the key features of communicative figurations by identifying communication forms, identifying a media ensemble, a constellation of actors and a framing (of family). In this sense, although locally dispersed and mediated, this communicative figuration constitutes a social practice that is grounded in tradition, enables the maintenance of a coherent sense of social identity and affords the expression of capabilities. In a certain sense, this example represents the transformation of the communicative figuration of the traditional family to that of the transnational migrant family. This media-saturated social practice also illustrates the way that communicative figurations can provide a context for an ethical form of life grounded in multi-optionality, mediacy, and an asynchronicity or asymmetry of roles.

CONCLUSION

The argument formulated at the beginning of this chapter is that we cannot understand present forms of mediatization and their relation to "the good life" by focusing on only one single medium—we must consider the complexity of the present media environment. Following the process sociology of Norbert Elias brings a focus on the transforming communicative figurations people are embedded in within their everyday lives. In relation to questions of "the good life," this understanding of media and social transformation maps well onto an extended media ethics inquiry, which is based on a broader sense of human needs.

Our examples of homeless people and transnational migrant families demonstrate a range of communicative figurations and their challenges. Homeless people are confronted with the situation that the media ensemble of more and more communicative figurations is based on new technologies. Consequently, there is a high risk of their being excluded from many relevant communicative figurations of contemporary life. In the meantime, the availability of mobile, digital media is increasingly important for them to address their "economic," "recognition" and "social needs." Therefore, access to the mobile phone and its multi-optionality to communicate becomes an important potential means of overcoming the exclusions of being homeless in a mediatized world. The use of a communication technology such as mobile phones offers them the chance to be involved in various communicative figurations that potentially overcome aspects of the ontological insecurity of living on the street.

The communicative figurations of everyday living have also changed for transnational migrant families. They engage with the strategic use of an ensemble of digital media—polymedia. The potential to select different combinations of media to articulate social relationships within the family at a distance results in new possibilities, but also new forms of complexity of daily life. In the best case, polymedia communicative figurations address the "social needs" of the migrant mothers and their children to sustain family relationships. Simultaneously, the new complexities become part of these social relationships, which are not caused by the migration as such, but by the affordances of the different combinations of new media.

These two cases of contemporary uses of new media were presented as examples of a mediatization process on a long-term scale. Because homeless people and migrants live in more or less precarious circumstances, having access to the media becomes important for them. The main point here is that the generally increasing multi-optionality of communication provides a variety of options for them to handle their lives communicatively—in the case of homeless people—by constructing a degree of ontological security; in the case of migrant families, by appropriating the options of polymedia for sustaining social relationships. In addition, the increasing mediacy of communication is important for them, creating a challenge for homeless people who struggle to access the mediated chains of communicative action, and allowing migrant families to sustain their social relations across a long distance. In these differences the two cases demonstrate altogether the asynchronicity in which mediatization takes place: being homeless in a mediatized world is a more open situation, and might result in the feeling of a far-reaching exclusion and, we have suggested, requires a more expanded ethical commitment of a broader community of users to create a viable communicative figuration that would include the homeless. Living as a migrant in times of polymedia is a mediatized experience that offers new possibilities for actively addressing "social needs."

These are the ambivalences surrounding mediatization. On the one hand, mediatization offers opportunities for addressing the various needs of our everyday lives, even in rather precarious situations. On the other hand, it can also be a burden as it might mean that people cannot address their needs without having access to certain media. This is the reason why mediatization as such is not simply good or bad. If we consider mediatization normatively, we have to frame it in a kind of extended media ethics that moves the virtues of daily life to the foreground. Only in this way can we offer arguments about the possible role of media for "the good life"—especially through the perspective of communicative figurations.

REFERENCES

Brueckner, M., Green, M., & Saggers, S. (2011). The trappings of home: Young homeless people's transitions towards independent living. *Housing Studies*, *26*(1), 1–16.

Burkum, K. (1999). Homelessness, virtue theory, and the creation of community. In G. J. M. Abbarno (Ed.), *The ethics of homelessness: Philosophical perspectives* (pp. 79–92). Value inquiry book series. Amsterdam: Rodopi.

Burzan, N., Lökenhoff, B., Schimank, U., & Schöneck, N. (2008). *Das Publikum der Gesellschaft. Inklusionsverhältnisse und Inklusionsprofile in Deutschland* [The audience of the society. Inclusion relations and inclusion profiles in Germany]. Wiesbaden: VS.

Couldry, N. (2004). Theorizing media as practice. *Social Semiotics*, *14*(2), 115–132.

Couldry, N. (2012). *Media, society, world: Social theory and digital media practice*. Cambridge: Polity Press.

Couldry, N. (2013). Living well with and through media. In N. Couldry, M. Madianou, & A. Pinchevski (Eds.), *Ethics of media* (pp. 39–56). New York: Palgrave Macmillan.

Couldry, N., & Hepp, A. (2012). Media cultures in a global age: A transcultural approach to an expanded spectrum. In I. Volkmer (Ed.), *Handbook of global media research* (pp. 92–109). Malden, MA: Wiley-Blackwell.

Couldry, N., & Hepp, A. (2013). Conceptualizing mediatization: Contexts, traditions, arguments. *Communication Theory*, 23(3), 191–202.

Elias, N. (1978). *What is sociology?* London: Hutchinson.

Elias, N. (2000). *The civilizing process*. Oxford: Wiley-Blackwell. (Original work published 1939)

Elias, N. (2010). *The society of individuals*. Dublin: UCD Press. (Original work published 1991)

Elias, N., & Scotson, J. L. (1994). *The established and the outsiders: A sociological enquiry into community problems* (2nd ed.). London: Sage.

Ethical Pioneer (2014, June 6). # Anti homeless floor studs. So much for community spirit: (. [PHOTO: Anti-homeless floor studs.]. [Tweet]. Retrieved from https://twitter.com/ethicalpioneer/statuses/474981723022049280

Georgiou, M. (2006). *Diaspora, identity and the media: Diasporic transnationalism and mediated spatialities*. Cresskill, NJ: Hampton Press.

Goffman, E. (1974). *Frame analysis: An essay on the organization of experience*. Cambridge, MA: Harvard University Press.

Hartmann, M. (2014). Home is where the heart is? Ontological security and the mediatization of homelessness. In K. Lundby (Ed.), *Mediatization of communication* (pp. 641–660). Berlin: de Gruyter.

Hasebrink, U. & Hölig, S. (2014). Topografie der Öffentlichkeit [Topography of the public sphere]. *APuZ*, 22–23/2014, 16–22.

Hepp, A. (2013a). *Cultures of mediatization*. Cambridge: Polity Press.

Hepp, A. (2013b). The communicative figurations of mediatized worlds: Mediatization research in times of the 'mediation of everything'. *European Journal of Communication*, 28(6), 615–629.

Hepp, A., & Hasebrink, U. (2014). Human interaction and communicative figurations. The transformation of mediatized cultures and societies. In K. Lundby (Ed.), *Mediatization of communication* (pp. 249–272). Berlin: de Gruyter.

Hjarvard, S. (2013). *The mediatization of culture and society*. New York: Routledge.

Horst, H., & Miller, D. (Eds.). (2012). *Digital anthropology*. London: Berg.

Hutchby, I. (2001). Technologies, texts and affordances. *Sociology*, 35(2), 441–456.

Jensen, K. B. (2013). Definitive and sensitizing conceptualizations of mediatization. *Communication Theory*, 23(3), 203–222.

Knoblauch, H. (2013). Communicative constructivism and mediatization. *Communication Theory*, 23(3), 297–315.

Krotz, F. (2009). Mediatization: A concept with which to grasp media and societal change. In K. Lundby (Ed.), *Mediatization: Concept, changes, consequences* (pp. 21–40). New York: Lang.

Lahire, B. (2011). *The plural actor*. (D. Fernbach, Trans.). Cambridge: Polity Press. (Original work published 2001)

Laing, R. D. (1960). *The divided self: An existential study in sanity and madness*. London: Tavistock.

Livingstone, S. M. (2009). On the mediation of everything. *Journal of Communication*, 59(1), 1–18.

Livingstone, S. M., & Lunt, P. (2014). Mediatization: An emerging paradigm for media and communication research? In K. Lundby (Ed.), *Mediatization of communication* (pp. 703–724). Berlin: de Gruyter.

Luhmann, N. (2012). *Theory of society* (Vol. 1). Stanford, CA: Stanford University Press.

Lundby, K. (Ed.). (2014). *Mediatization of communication*. Berlin: de Gruyter.

Madianou, M. (2014). Polymedia communication and mediatized migration: An ethnographic approach. In K. Lundby (Ed.), *Mediatization of communication* (pp. 323–348). Berlin: de Gruyter.

Madianou, M., & Miller, D. (2012). *Migration and new media: Transnational families and polymedia*. Milton Park, UK: Routledge.

Madianou, M., & Miller, D. (2013). Polymedia: Towards a new theory of digital media in inter-personal communication. *International Journal of Cultural Studies, 16*(2), 169–187. doi:10. 1177/1367877912452486

Pollio, D. E., Batey, D. S., Bender, K., Ferguson, K., & Thompson, S. (2013). Technology use among emerging adult homeless in two U.S. cities. *Social Work, 58*(2), 173–175.

Rice, E., Monro, W., Barman-Adhikari, A., & Young, S. D. (2010). Internet use, social networking, and homeless adolescents' HIV/AIDS risk. *Journal of Adolescent Health, 47*(6), 610–613.

Roberson, J., & Nardi, B. (2010). Survival needs and social inclusion: Technology use among the homeless. Proceedings of the *2010 ACM Conference on computer-supported cooperative work* (pp. 445–448). Retrieved from http://research.microsoft.com/en-us/um/redmond/groups/connect/CSCW_10/docs/p445.pdf

Rosa, H. (2013). *Social acceleration: A new theory of modernity*. (J. Trejo-Mathys, Trans.). New York: Columbia University Press.

Schütz, A. (1967). *The phenomenology of the social world*. Evanston, IL: Northwestern University Press.

Sen, A. (1992). *Inequality reexamined*. Oxford: Clarendon.

Sen, A. (1999). *Development as freedom*. Oxford: Oxford University Press.

Siapera, E. (2005). Minority activism on the web: Between deliberative democracy and multicultural-ism. *Journal of Ethnic and Migration Studies, 31*(3), 499–519.

Silverstone, R. (2006). *Media and morality: On the rise of the mediapolis*. Cambridge: Polity Press.

Simmel, G. (1971). The metropolis and mental life. In G. Simmel, *On individuality and social forms* (pp. 324–339). Chicago, IL: University of Chicago Press. (Original work published 1903)

Spiegel. (2000, August 2). Grünen-Politiker fordert: Handys für Obdachlose. *Spiegel Online*. Retrieved from www.spiegel.de/politik/deutschland/gruenen-politiker-fordert-handys-fuer obdachlose-a-87615.html

Sreberny, A. (2005). 'Not only, but also': mixedness and media. *Journal of Ethnic and Migration Studies, 31*(3), 443–459.

Tomas, A., & Dittmar, H. (1995). The experience of homeless women: An exploration of housing histories and the meaning of home. *Housing Studies, 10*(4), 493–515.

Tomlinson, J. (2007). *The culture of speed: The coming of immediacy*. New Delhi: Sage.

Wajcman, J. (2002). Addressing technological change: The challenge to social theory. *Current Sociology, 50*(3), 347–363.

Winnicott, D. W. (1971). *Playing and reality*. London: Tavistock.

Woelfer, J. P., & Hendry, D. G. (2012). Homeless young people on social network sites. Proceedings from *Conference on Human Factors in Computing Systems held in Austin, TX, 5–10 May 2012*. Retrieved from http://dub.washington.edu/djangosite/media/papers/WoelferHendry_HYPSNS_CHI2012.pdf

CHAPTER TWELVE

Reimagining the Good Life with Disability

Communication, New Technology, and Humane Connections

MERYL ALPER, UNIVERSITY OF SOUTHERN CALIFORNIA, USA
ELIZABETH ELLCESSOR, INDIANA UNIVERSITY, USA
KATIE ELLIS, CURTIN UNIVERSITY, AUSTRALIA
GERARD MICHAEL GOGGIN, UNIVERSITY OF SYDNEY, AUSTRALIA

Many deeply cherished notions of "the good life" are based on limiting notions of humans, things, and their environment. In particular, "the good life" is often imagined as a realm beyond illness, impairment, and especially, disability. This view is informed by deficit models of disability, which individualize disability rather than explore the "socio-cultural conditions of disablism" (Goodley, 2011, p. 29). With contemporary communication and new media, disability is even more seen as an impediment, barrier, or tragedy, to be overcome with digital technology. Regrettably, the widely shared experience of disability and its complex relationships with communication are only rarely seen as a resource for how we achieve "the good life," in our own lives and societies, now and in the future.

Indeed while the field of communication increasingly engages with questions concerning marginalized populations—including issues of race, ethnicity, gender, sexuality, and diasporic populations—the study of disability and its relationship with social, cultural, and political life is stagnant. New media are often hailed as a great "equalizer" for people with disabilities. Such arguments though tend to obscure the complex ways in which disability and technology intersect for better and for worse in the lives of people with various disabilities from diverse backgrounds. However, with the rise of new social movements, disciplinary formations, and theories—such as critical disability studies—communication studies is slowly engaging with the challenges and new conceptual possibilities disability offers.

Accordingly in this chapter, we take up pressing yet sorely neglected questions of disability and communication in order to illuminate how we might see "the good life" in much more enabling, humane, and democratic ways. To do so, first, we discuss the state of the art of communications and disability theory. Second, we focus on the ways in which contemporary information and communication technologies (ICTs) shape and are shaped by notions of disability and ability. Third, we identify and debate the lessons from disability and communication studies that help us to rethink "the good life," especially in the new media environment.

THEORETICAL APPROACHES: COMMUNICATION
AS IF DISABILITY MATTERED

Communication, as a discipline, has much to gain from taking on the theoretical and philosophical questions raised by the notion of leading a good life with disability. That is, rather than visions of "the good life" being premised on the assumption that impairment and disability need to be overcome, transcended, or cured, there can flow new ways of imagining, creating, and inhabiting lives—that embrace rather than reject or disavow disability.

Thus, disability studies scholarship calls into question the naturalness of "normal" as a category (Davis, 2002), and, as a corollary, the idea that there is a singular "good life" to be had by all. Societal determination of which individuals have marked disabilities and whose needs are "special" greatly varies across history, geography, and cultures (Garland-Thomson, 1997; Moser, 2000; Goodley, 2011). Critical approaches to disability draw attention to the ways in which notions of "the good life" provide a basis for government and society to discriminate against and segregate individuals with disabilities from their fellow citizens (Tremain, 2005). Disability is part of, and very much interacting with, other discourses of "otherness"—such as race, gender, and sexuality—in the production of bodily differences often seen as a social stigma and grounds for exclusion and discrimination (Anspach, 1979; Goffman, 1963; Kafer, 2013; Siebers, 2008). The attainment of individual "well-being" is a socially constructed ideal, one that rarely acknowledges "being unwell" as a state of wholeness or completion (McRuer, 2006).

Communication and its scholarship have a profound role to play in reimagining "the good life" via disability. After all, the basic constituents of communication—speaking, hearing, listening, and writing—all stand to be altered if disability is introduced. This encompasses formulations of language and signs, image and symbols, economies and production of communication. Take, for instance, the question of speaking—a long-standing topic in communication as much as culture generally. For a range of different groups, speaking—or articulation and expression more broadly conceived—occurs quite differently than norms of "speech communication" would suggest. For the most part, communication scholars are comfortable with the idea that communication is a very broad arena indeed—and that communication is multi-modal, that most communication involves forms of mediation, and, more recently, that new media do not simply concern screen-based media. When it comes to disability, however, there has been little reflection on what its complex entanglement of meanings, practices, and media has to say for our ideas of communication.

The magnitude of this enterprise can be indicated by reflecting on Peters' (1999) resonant argument in relation to the idea of communication: "Communication as a person-to-person activity became thinkable only in the shadow of mediated communication" (p. 6). In relation to disability, this would at the least imply that we need to think about the history of mediated communication forms to understand disability and communication. More radically still, we could formulate the bold hypothesis that communication only became thinkable in the penumbra of *disability*-mediated communication. Or, rather, we need to think through the history of disability and communication to understand communication as a general concept. In the brief compass of this chapter, we can only offer this provocation—suggesting that a future priority for research and debate is the argument that our bedrock models of communication still are not centrally informed by the materialities and imaginaries of disability communication.

Consider, for instance, Deaf people "speaking" via sign language (Bauman, 2008). Or people with profound speech disabilities communicating via gestures, signs, keyboards, and now tablet computers. Then there is the question of listening, an emerging topic of research (Dobson, 2014; Lacey, 2013). Questions of voice, who speaks, from what position, and in what context and power relations have been central to communication and media studies (Couldry, 2010). Yet questions of listening—who listens to whom, why, for what purposes, and with what implications—have not often been systemically posed (O'Donnell, Lloyd, & Dreher, 2009). From a critical disability standpoint, listening suggests many important challenges to how we understand communication, its architectures, imperatives, and politics (Goggin, 2009). To attend adequately to people who communicate "differently"—that is, the non-dominant, non-normative ways that might be associated with autism culture, blind culture, people with intellectual disability, or people who take longer than some others to articulate their thoughts, or who use different words, sounds, or signs—requires a shift in how we understand communication.

Despite the fertile promise of communication and disability, it truly has been a specialized, minority pursuit in communication and media studies. It has been the province of particular professions and researchers focusing on improving communication by and with people with disabilities (cf. Abudarham & Hurd, 2002; Parr, Duchan, & Pound, 2003). It certainly has not been seen as a *transformative* force in communication, as gender, sexuality, race, post-colonialism, and other intellectual and social movements are commonly regarded. Consider, for instance, the paucity of full-length studies of communication and disability, or the fact that there is just *one* reference work on communication and disability, now 15 years old (viz. Braithwaite & Thompson, 2000). While there are various anthropologists who have contributed important research—such as Boellstorff (2008), Ginsburg (2012), and Rapp (Ginsburg & Rapp, 2013)—there are very few scholarly books dedicated to media and disability *per se* with a notable exception of Haller (2010). By contrast, disability has been much more visible in cinema and film studies, and literary studies (Garland-Thomson, 1997; Longmore, 1985; Norden, 1994). Happily, a wave of scholarship and publications is now emerging in disability and media (e.g., Ellis, 2015; Ellis & Goggin, 2015; Rodan, Ellis, & Lebeck, 2014).

So, our argument in this chapter is two-fold. First, to think about and discuss "the good life," we need to consider the fundamental challenges disability poses— and indeed embrace the opportunities it offers for doing things otherwise, perhaps much better, fairer, sustainably, and, even more happily. Second, communication as a field needs to engage with disability, as a long overdue priority. To substantiate this argument, we provide a concrete set of examples revolving around technology. A great deal of the 2014 Seattle ICA conference, highlighted in its call for papers, plenaries, and many of its panels and presentations, interpreted the theme of "the good life" via the question of technology. So the three examples that follow in the second half of this chapter provide a common line of inquiry, analysis, and conceptualization concerning communication and technology—from a critical disability perspective. In doing so, we make common cause between disability studies and science and technology studies (STS).

THE INTERDEPENDENCE OF DIS/ABILITY AND ICTS

A dominant way of seeing disability in relation to communication and technology is embodied in the concept of "assistive technology." That is, technology associated with disability is typically positioned as "assistive" and "specialized," rather than mainstream. Placing communication scholarship in conversation with disability studies and with STS contributes to a larger public dialogue about the tools that give assistance to all human bodies, as opposed to conceiving of "assistive technologies" as outside the domain of communication studies (Goggin & Newell, 2003). Some cultural anthropologists have argued that all human communication is "assisted" in some manner by learned techniques, linguistic equipment, and other resources (Moser & Law, 2003; Reno, 2012). What makes an ICT "assistive" is largely shaped by the context of its use as opposed to the purely intrinsic qualities of the technology or the technology user (Mills, 2011).

STS offers ways to understand ICTs as having a politics in that they contribute to the production of particular social orders and the perpetuation of social ideals (Winner, 1986). One pervasive ideology in technology circles is the idea that "bits" or the digital will eventually replace "atoms" or the physical (Negroponte, 1996). McLuhan (1964) theorized media as the ultimate bodily extension while also warning of its potential to diminish human capacities, a tension evidenced throughout modern history (Foster, 2004). Critical STS scholars have voiced concern with this conception of prosthesis, as well as the romanticization of the "cyborg" figure (Haraway, 1991; Hayles, 1999) and its celebration of the merging of organic and mechanic parts. Kafer (2013), Mills (2011), Sobchack (2004) and others call for prosthesis to be understood as more than a metaphor but as a material reality.

Disability also figures heavily into the aesthetics of media, art, and architecture (Siebers, 2010). For example, "good" web design and "accessible" web design are often framed in opposition to one another (Ellcessor, in press). Viewing disability in terms of "accommodation," the term often used in anti-discrimination law (cf. Pothier & Devlin, 2006; Pullin, 2009), instead figures disability as resource and inspiration for creativity and innovation. Williamson (2012) explores how today's "do-it-yourself" or "Maker" movement among (often white and middle-class) U.S. consumers is rooted in post-war society. During that time, more people were living with significant physical impairments than ever before (e.g., surviving polio or loss of limbs in WWII). Faced with built-in obstacles to living in their own homes and sharing public spaces, a privileged disabled population (also generally white and middle-class) re-designed their environments as a form of self-preservation at the margins of society. The design of assistive technologies is always historically situated within the consumer culture of a given era and how dreams of "the good life" are packaged and sold.

Against this background, the following examples highlight various ways in which disability is bound up with conceptions of technological and social progress. We focus here on three technologies—tablet computers, closed captioning, and video descriptions—often figured by state and corporate actors as inherently improving the lives of individuals with disabilities. Each case study illuminates the cultural, historical, political, and economic forces that for better and for worse shape the lived experience of disability and so too the communicative technoscape through which disability is lived in contemporary Western society.

Tablet Computing

Recent developments in mobile technology have had a dramatic impact on individuals with disabilities, particularly those who have significant difficulty speaking. People across the lifespan (most famously, physicist Stephen Hawking) use speech-generating or augmentative and alternative communication (AAC) devices because they desire additional means for expression. While computerized AAC devices have traditionally cost thousands of dollars, less expensive tablet computers equipped with specialized software applications (or "apps") for AAC are rapidly overtaking the market (McNaughton & Light, 2013). These devices also enable other forms of mediated communication, such as email and social networking sites.

How are tablets, used primarily though not exclusively as AAC devices, similar or different from any other tablet computer a person might obtain? It is a complex question, but one entry point is through analyzing how well leading social scientific theories predicting ICT adoption explain the purchase and use of tablets for AAC and other purposes. Any model of technology adoption that counts people with disabilities as rightful technology users must take into account the variation of lived experiences with disability.

Two leading theories—Diffusion of Innovations (DOI) theory (Rogers, 2003) and the Model of Adoption of Technology in Households (MATH) (Brown & Venkatesh, 2005; Venkatesh & Brown, 2001)—are particularly relevant starting points. Researchers studying technology use by people with disabilities often employ the DOI framework (e.g., Hurst & Tobias, 2011). MATH attempts to explain the process by which households make purchase decisions about consumer electronics. Family members play a significant role in the adoption of tablet-based AAC devices. Recent surveys have found that 68% to 73% of individuals using an iPad for AAC obtained it through a family purchase (e.g., McBride, 2011).

DOI and MATH have particular advantages and drawbacks. Strengths of the theories encompass three key areas: discontinuance, confirmation, and contingency. First, these theories account for when a person adopts a technology and then ceases to use the device. It is estimated that 8% to 75% of all assistive technologies are adopted but their use is eventually discontinued, depending on the device (Scherer, 2005). A person might stop using an AAC device because they do not know what to do or whom to contact when a device needs repair or replacement. Second, confirmation, or social reinforcement for an adoption decision, is especially important in the digital age. Hurst and Tobias (2011) note that for users with disabilities, having access to online communities of fellow assistive technology users shapes adoption. Third, these theories also account for contingency, or when people reference earlier purchase decisions when making new technology adoption decisions. The tablet-based AAC devices that people with disabilities use may impact their adoption and use of other ICTs and vice versa.

There are also weaknesses to these models. They do not account for the bureaucracy in assistive technology delivery, which varies by country. While some AAC users can afford to buy tablet devices out-of-pocket, others must go through a time and labor-intensive process of obtaining outside funding. In the U.S., Medicare,

Medicaid, and private health insurance plans generally do not fund tablets fearing the fragility of technologies like the iPad and the potential for fraud. Also, Rogers (2003) proposes that the observability of a technology (i.e., the visibility of its use) increases adoption rates. This does not necessarily hold true for assistive technologies and can in fact have the opposite effect. Some devices are discarded because they draw unwanted attention to users (Shinohara & Wobbrock, 2011). Observability, as a predictor of technology adoption, is complicated by the identity politics surrounding disability, design, and assistive technologies.

In sum, existing models of ICT adoption are not a perfect match for explaining the processes by which people with complex communication needs adopt tablet-based computers for AAC. Certain concepts are useful and relevant, but there are complications in directly applying these theories to commercially available tablet devices and AAC apps. They are problematic in light of the political economy of production and delivery systems for tablet-based AAC devices, as well as the values and meanings associated with technology used by individuals with disabilities in specific sociocultural and historical contexts.

Goggin and Newell (2007), in their writing on the "business of digital disability," argue that to bring about truly accessible ICTs for all requires revisiting "what it is we mean by inclusion and how too often we leave un-reformed the exclusionary power relations and technologies that require inclusion in the first place" (p. 166). More critical thought is needed regarding how it is that people with communication disabilities—many whose participation in the social and public arena and attainment of "the good life" is intimately tied to mobile devices—are so understudied among communication scholars in the first place.

Closed Captioning

A second provocation regarding the accessibility of "the good life" comes from the shifts that digitization has brought to closed captioning for digital television and online video. First, the "closed" component of captioning indicates that it is an opt-in service, hidden by default, which reinforces the invisibility of disability in much of mainstream culture and in utopian visions of a technological future (Ellcessor, 2012; Kafer, 2013). Moreover, the rise of mobile and streaming forms of media viewing and concurrent legislation in the U.S. has created a context in which this invisibility is perpetuated and extended. Ideologies of "disability as deficit" are reinforced through the deskilling of labor and denial of the cultural work of captioning, which may produce low-quality captions that poorly serve d/Deaf and hearing-impaired audiences.

The technical and cultural forms of skilled labor that are part of creating captions are, like captions themselves, often invisible. Downey's (2008) history of closed captioning for television demonstrates the invisibility of the geographically dispersed stenographers who create captions, and ties them to processes of digitization and related professions such as courtroom reporting or real-time captioning for d/Deaf and hard of hearing individuals. These workers often work in non-urban settings, or even from home, and are predominantly female. Their work is not understood to be creative or transformative of content. However, captioners are engaged in "accessibility labor." Downey (2008) understands accessibility labor as a form of translation that allows information to be accessed and circulated, and notes that, "Real-time stenographers on television must adapt scripts and speedy verbal delivery to the very limited time and space of the television screen, taking care to judiciously drop words, phrases, and even sentences when the flow becomes too fast and when the commercial break looms" (p. 158).

The processes described here indicate that there is a cultural awareness inherent to the production of high-quality captions, as captioners must have the literacy and dexterity to make moment-by-moment decisions and produce a final product that is different from but equivalent to the original. The choices made by captioners effectively alter content to make it accessible, resulting in the "dilemma of accuracy": difficulty in finding a balance between fidelity to the spoken components and its stylistic or larger function within a context of restricted space and limited time in which viewers could read content synced with the video (De Linde & Kay, 1999).

Although this labor has always been invisible, and undervalued, digitization has entrenched an ideological inferiority of accessible media and its users by allowing this work to be further deskilled or made unnecessary. Downey (2014) suggests that information labor such as captioning might be "entirely extracted from human minds and hands existing as automated algorithmic labor" (p. 145). In one sense, if a literal transcoding from audiovisual to textual content is possible, it would seem to be a boon to access. However, the drive to automate captioning labor coincides with the rise of industries such as search engines and social networking sites that need vast quantities of textual data upon which to build their systems (Ellcessor, 2012), and the increase in U.S. legal requirements for captioning. Thus, automated systems like Google's auto-captions are rarely highly accurate.

Such automation is a logical outcome of what Striphas refers to as algorithmic culture, in which "computers, running complex mathematical formulae, engage in what's often considered to be the traditional work of culture: the sorting, classifying, and hierarchizing of people, places, objects, and ideas" (Granieri & Striphas, 2014, n.p.). In this environment, textual data have a heightened market value as it can be used to create and improve algorithms; its value as an assistive technology and its cultural value as a translation for viewers with disabilities are correspondingly deprioritized in accordance with a deficit model of disability in which disabled audiences are less deserving of access to mediated communication than are mainstream audiences, advertisers, or even web crawlers.

Digitization, automation, and the proliferation of available forms of mediated communication are often invoked as a signal of "the good life," or an impending utopian future in which information is increasingly accessible, transformable, personalized, optimized, and circulated. Looking to the ways in which these same processes eviscerate the cultural values and possible political equity conveyed through assistive forms of media communication, however, indicates that this futurism is a fantasy of dominant classes. In imagining such a future, it is often suggested that problems of disability, inequality, and prejudice might have technological solutions; these solutions inevitably take the form of cure, meritocracy, and tolerance, failing to interrogate larger social structures and ideological investments (Kafer, 2013; Nakamura, 2002). Automation of captioning labor and the market value of textual data may appear as a rationalizing force for increasing available information, but they simultaneously may lower the quality of captions as an assistive technology and further marginalize audiences with disabilities by providing them only a second-class form of access to media content.

Video Description

Our third example for understanding disability communication and access to "the good life" is video description. It is an audio feature that describes important visual elements of a television show, movie, or performance between lines of dialogue and allows people with vision impairment access to a predominantly visual medium. It outlines relevant facial expressions, visual jokes, costuming and anything else essential to the story that a person with vision impairment could miss out on.

In particular, the availability of video description on television is increasingly recognized as vital to the social inclusion of people with vision impairments (American Foundation for the Blind, 1997; Cronin & King, 1998; Ellis, 2014). As an American viewer expressed in the 1980s,

> …. [It was] was very emotional. I found myself pacing the floor in tearful disbelief. It was like somebody had opened a door into a new world, in which I was able to see with my ears what most people see with their eyes. (Cronin & King, 1998, n.p.)

More recently, a 12-week technical trial conducted on the Australian public broadcaster ABC found viewers celebrating a renewed experience of social inclusion (Madson, 2013; Mikul, 2013) with video description enabling them to "watch" television independently in their own homes:

> To be able to come home at the end of the day, sit on my lounge chair, turn on the TV and…be able to relax watching that. Not having to squint or try to figure out what's going on or pester other people to say, "Hey, what's just happened there?" It was a wonderful thing. (K. Ellis, personal communication, May 13, 2014)

People with vision impairment frequently connect the availability of video description to notions of "the good life" because it offers access to information and recreation, independence, and the feeling of social inclusion. As Lauren Henley, National Policy Officer at Blind Citizens Australia, explains,

> You might think that missing out on television is no great loss, but it's about more than watching the latest episode of *Days of our Lives*. Like the rest of my friends and family, I want to have choice about what I watch and have the ability to be informed about what is going on in the world. I lost many things when I lost my sight, but one of the things that I lost was social inclusion. (Henley, 2012)

Video description has progressed with the digitization of television (Ellis, 2014; Media Access Australia, 2012; Utray, de Castro, Moreno, & Ruiz-Mezcua, 2012). Despite the potential to create "new worlds" and new forms of communication for people with disabilities, digital technologies are not inherently accessible (Ellis & Kent, 2011). Different countries have taken different approaches to its implementation. In the U.S., the Federal Communications Commission (FCC) directed "the four big TV networks and the five biggest cable networks to show 50 hours of audio described programmes per quarter by April 2002" (Mikul, 2010, p. 5). The Motion Picture Association of America challenged this and the Supreme Court eventually ruled in their favor. However, in the meantime, the networks had already begun to comply with the FCC's mandate. In 2010, the Twenty-First Century Communications and Video Accessibility Act restored the rules earlier set up by the FCC to mandate four hours of audio described content per week (Media Access Australia, 2012).

While the Australian government acknowledged the potential of video description for digital and online television in a 2010 policy discussion paper, there is still no specific legislation in place, and as a result, no Australian broadcasters offer this feature (Ellis, 2014). When funding was not continued for the ABC service, a coalition of viewers with vision impairment initiated legal action against the Australian government and ABC under Australia's Disability Discrimination Act. The case is currently undergoing mediation (Ellis, 2014).

A recent report by the European Union found that accessibility is more widely available on digital and online television in countries where legislation is in place (Kubitschke, Cullen, Dolphin, Laurin, & Cederbom, 2013). In 2012, OfCom, the independent regulator and competition authority for the U.K. communication industries, created *The Code on Television Access* to help encourage accessibility for the blind and vision impaired. Section 8 of this code stipulates video description targets up to a total of 10% of content after 5 years of broadcasting while still allowing for some exemptions if audience share is less than 0.05%, or where there are technical or financial difficulties (OfCom, 2012). Following the introduction of this code, broadcasters began exceeding their minimum requirements, with some achieving 100% (OfCom, 2013).

Taken together, three approaches can be distilled to video description accessibility as a potential contribution to, "the good life" for audiences with vision impairment: (1) industry innovation emphasised by governments and legislators, (2) the importance of legal interventions through disability discrimination acts by activists, and (3) industry regulators such as OfCom playing a vital role in not only ensuring the availability of video description but also in suggesting opportunities for mainstream benefits.

CONCLUSION: DISABILITY AND "THE GOOD LIFE" IN THE DIGITAL AGE

In this chapter, we presented three examples to illustrate the dynamics, tensions, and texture of communication in the lives of people with disabilities. The case of AAC devices and assistive technologies highlights a neglected area in theories of technology adoption and use, as well as media policy and political economy. It shows that the AAC area is no longer, if it ever was, a specialized realm where disability and technology professionals rule. Rather, especially in the smartphone and tablet phase of mobile, online, and social media, the social shaping of AAC is very much bound up with mainstream questions of design, production, and consumption of digital technology.

Closed captioning is another area rendered marginal, indeed invisible, in mainstream media and culture. Yet in many ways, as our analysis suggests, the political and cultural economies associated with the shaping of closed captioning raise in a vivid and pressing form key questions to do with digital media technology generally. These are questions of labor as well as cultural citizenship, and matters of participation, equality, and enjoyment of common societal resources—and whether these will be open to all, regardless of class, wealth, or ascribed, subjected difference such as disability.

Video description also is an area of considerable significance and impact in terms of the cultural access, participation, and resources afforded to people with disabilities. Although it is often denigrated as a "low" or "anti-social" cultural form, television has gained acceptance as a vitally important, internationally ubiquitous medium. If one experiences exclusion from television, as various scholars have argued, in subtle ways one also is marginalized and excluded from participation in the public and private sphere. Especially in the emerging television ecologies associated with post-broadcast, digital networked infrastructures, technologies, and participatory cultures, technologies such as video description are strategically important for ensuring contemporary diversity of audiences is acknowledged—and that all audience members are embraced.

It is no accident that these examples each have to do with technology. In its various guises, technology has moved to the center stage in contemporary societies. Technology provides crucial tools, infrastructures, support, and networks, as well as feeding into forms of social and cultural capital, that structure patterns of domination, participation, exclusion, and belonging. A longstanding insight from STS has been that technology amounts to "society made durable" (Latour, 1991, p. 103). In the cases of disability and communication technology we have explored here, this is clearly the case. However, there are more complex dimensions to communication technology that disability helps us to appreciate, which has also been the focus of developments in STS theory, as much as it has been in communication, cultural, media, and social theory. Returning to the argument we made earlier in this chapter, it is our view that a critical examination of disability has much to teach us about communication. Hopefully, the analysis provided here has indicated the fruitfulness—if not urgent necessity—of a disability turn in communication studies. Nowhere more does this stand to be useful than in confronting the question of technology, a cardinal matter in communication today—but especially in the politics of the future, that we predicate when we imagine and debate "the good life."

REFERENCES

Abudarham, S., & Hurd, A. (Eds.). (2002). *The management of communication needs in people with learning disability*. London: Whurr.

American Foundation for the Blind. (1997, August 25). *Who's watching? A profile of the blind and visually impaired audience for television and video*. Retrieved from http://www.afb.org/section.aspx?FolderID=3&SectionID=3&TopicID=135&DocumentID=1232#frustrating

Anspach, R. R. (1979). From stigma to identity politics: Political activism among the physically disabled and former mental patients. *Social Science & Medicine: Part A, 13*, 765–773.

Bauman, H.-D. L. (Ed.). (2008). *Open your eyes: Deaf studies talking*. Minneapolis, MN: University of Minnesota Press.

Boellstorff, T. (2008). *Coming of age in Second Life: An anthropologist explores the virtually human*. Princeton, NJ: Princeton University Press.

Braithwaite, D. O., & Thompson, T. L. (Eds.). (2000). *Handbook of communication and people with disabilities*. Mahwah, NJ: Erlbaum.

Brown, S. A., & Venkatesh, V. (2005). Model of adoption and technology in households: A baseline model test and extension incorporating household life cycle. *MIS Quarterly, 29*(3), 399–426.

Couldry, N. (2010). *Why voice matters: Culture and politics after neoliberalism*. Thousand Oaks, CA: Sage.

Cronin, B. J., & King, S. R. (1998, September). *The development of the descriptive video services*. Retrieved from http://www2.edc.org/NCIP/library/v&c/Cronin.htm

Davis, L. (2002). *Bending over backwards: Disability, dismodernism and other difficult positions.* New York: NYU Press.

De Linde, Z., & Kay, N. (1999). *The semiotics of subtitling.* Manchester: St. Jerome.

Dobson, A. (2014). *Listening for democracy: Recognition, representation, reconciliation.* Oxford: Oxford University Press.

Downey, G. J. (2008). *Closed captioning: Subtitling, stenography, and the digital convergence of text with television.* Baltimore: Johns Hopkins University Press.

Downey, G. J. (2014). Making media work: Time, space, identity, and labor in the analysis of information and communication infrastructures. In T. Gillespie, P. J. Boczkowski, & K. A. Foot (Eds.), *Media technologies: Essays on communication, materiality, and society* (pp. 141–166). Cambridge, MA: MIT Press.

Ellcessor, E. (2014). <ALT= "Textbooks">: Web accessibility myths as negotiated industrial lore. *Critical Studies in Media Communication, 31*(5), 448–463.

Ellcessor, E. (2012). Captions on, off, on TV, Online: Accessibility and search engine optimization in online closed captioning. *Television & New Media, 13*(4), 329–352.

Ellis, K. (2014). Digital television flexibility: A survey of Australians with disability. *Media International Australia, 150,* 96–105.

Ellis, K. (2015). *Disability and popular culture.* London: Ashgate.

Ellis, K., & Goggin, G. (2015). *Disability and the media.* New York: Palgrave Macmillan.

Ellis, K., & Kent, M. (2011). *Disability and new media.* New York: Routledge.

Foster, H. (2004). *Prosthetic gods.* Cambridge, MA: MIT Press.

Garland-Thomson, R. (1997). *Extraordinary bodies: Figuring physical disability in American culture and literature.* New York: Columbia University Press.

Ginsburg, F. (2012). Disability in the digital age. In H. Horst & D. Miller (Eds.), *Digital anthropology* (pp. 101–126). London: Berg.

Ginsburg, F., & Rapp, R. (2013). Disability worlds. *Annual Review of Anthropology, 42,* 53–68.

Goffman, E. (1963). *Stigma: Notes on the management of spoiled identity.* New York: Simon & Schuster.

Goggin, G. (2009). Disability and the ethics of listening. *Continuum: Journal of Media and Cultural Studies, 23*(4), 489–502.

Goggin, G., & Newell, C. (2003). *Digital disability: The social construction of disability in new media.* Lanham, MD: Rowman & Littlefield.

Goggin, G., & Newell, C. (2007). The business of digital disability. *The Information Society, 23*(3), 159–168.

Goodley, D. (2011). *Disability studies: An interdisciplinary introduction.* Los Angeles: Sage.

Granieri, G. (Interviewer) & Striphas, T. (Interviewee). (2014, April 30). Algorithmic culture. "Culture now has two audiences: people and machines": A conversation with Ted Striphas. *Medium.* Retrieved from https://medium.com/futurists-views/2bdaa404f643

Haller, B. (2010). *Representing disability in an ableist world: Essays on mass media.* Louisville, KY: Advocado Press.

Haraway, D. (1991). A cyborg manifesto: Science, technology, and socialist-feminism in the late twentieth century. In D. Haraway, *Simians, cyborgs and women: The reinvention of nature* (pp. 149–181). New York: Routledge.

Hayles, N. K. (1999). *How we became posthuman.* Chicago, IL: University of Chicago Press.

Henley, L. (2012). *How audio-described TV has changed my world.* Retrieved from http://www.human-rights.gov.au/how-audio-described-tv-has-changed-my-world

Hurst, A., & Tobias, J. (2011). Empowering individuals with do-it-yourself assistive technology. In *Proceedings of the 13th International ACM SIGACCESS Conference on Computers and Accessibility* (pp. 11–18). New York: ACM.

Kafer, A. (2013). *Feminist, queer, crip*. Bloomington, IN: Indiana University Press.

Kubitschke, L., Cullen, K., Dolphin, C., Laurin, S., & Cederbom, A. (2013). *Study on assessing and promoting e-accessibility*. Retrieved from http://ec.europa.eu/digital-agenda/en/news/study-assessing-and-promoting-e-accessibility

Lacey, K. (2013). *Listening publics: The politics and experience of listening in the media age*. Cambridge: Polity.

Latour, B. (1991). Technology is society made durable. In J. Law (Ed.), *A sociology of monsters* (pp. 103–131). London: Routledge.

Longmore, P. K. (1985). Screening stereotypes: Images of disabled people in television and motion pictures. *Social Policy, 16*(1), 31–37.

Madson, G. (2013, 10 July). What just happened? Our audio description has vanished! *ABC.au*. Retrieved from http://www.abc.net.au/rampup/articles/2013/07/10/3800109.htm

McBride, D. (2011). AAC evaluations and new mobile technologies: Asking and answering the right questions. *Perspectives on Augmentative and Alternative Communication, 20*(1), 9–16.

McLuhan, M. (1964). *Understanding media: The extensions of man*. New York: McGraw-Hill.

McNaughton, D., & Light, J. (2013). The iPad and mobile technology revolution: Benefits and challenges for individuals who require augmentative and alternative communication. *Augmentative and Alternative Communication, 29*(2), 107–116.

McRuer, R. (2006). *Crip theory: Cultural signs of queerness and disability*. New York: NYU Press.

Media Access Australia. (2012). *Audio description on TV in the US*. Retrieved from http://www.media-access.org.au/television/audio-description-on-tv/audio-description-on-tv-in-the-us

Mikul, C. (2010). *Audio description background paper*. Ultimo: Media Access Australia.

Mikul, C. (2013, November 12). *Audio description—where to now?* Retrieved from http://mediaaccess.org.au/sites/default/files/files/AD_where_to_now.pdf

Mills, M. (2011). Deafening: Noise and the engineering of communication in the telephone system. *Grey Room, 43*, 118–143.

Moser, I. (2000). Against normalization: Subverting norms of ability and disability. *Science as Culture, 9*(2), 201–240.

Moser, I., & Law, J. (2003). "Making voices": New media technologies, disabilities, and articulation. In G. Liestøl, A. Morrison, & T. Rasmussen (Eds.), *Digital media revisited: Theoretical and conceptual innovation in digital domains* (pp. 491–520). Cambridge, MA: MIT Press.

Nakamura, L. (2002). *Cybertypes: Race, ethnicity, and identity on the Internet*. New York: Routledge.

Negroponte, N. (1996). *Being digital*. New York: Vintage.

Norden, M. F. (1994). *The cinema of isolation: A history of physical disability in the movies*. New Brunswick, NJ: Rutgers University Press.

O'Donnell, P., Lloyd, J., & Dreher, T. (2009). Listening, pathbuilding and continuations: A research agenda for the analysis of listening. *Continuum, 23*(4), 423–439.

OfCom. (2012, December 18). *Ofcom's code on television access services*. Retrieved from http://stakeholders.ofcom.org.uk/broadcasting/broadcast-codes/tv-access-services/code-tv-access-services-2013/

OfCom. (2013, September 25). *Television access services: Report for the first six months of 2013*. Retrieved from http://stakeholders.ofcom.org.uk/market-data-research/market-data/tv-sector-data/tv-access-services-reports/access-q1q2–13

Parr, S., Duchan, J., & Pound, C. (Eds.) (2003). *Aphasia inside out: Reflections on communication disability*. Maidenhead, Berkshire, UK: Open University Press.

Peters, J. D. (1999). *Speaking into the air: A history of the idea of communication*. Chicago: University of Chicago Press.

Pothier, D., & Devlin, R. (Eds.). (2006). *Critical disability theory: Essays in Philosophy, Politics, Policy, and Law*. Vancouver: University of British Columbia Press.

Pullin, G. (2009). *Design meets disability*. Cambridge, MA: MIT Press.

Reno, J. (2012). Technically speaking: On equipping and evaluating "unnatural" language learners. *American Anthropologist, 114*(3), 406–419.

Rodan, D., Ellis, K., & Lebeck, P. (2014). *Disability, obesity and ageing: Popular media identifications*. London: Ashgate.

Rogers, E. M. (2003). *Diffusion of innovations* (5th ed.). New York: Free Press.

Scherer, M. J. (2005). *Living in the state of stuck: How assistive technology impacts the lives of people with disabilities* (4th ed.). Cambridge, MA: Brookline Books.

Shinohara, K., & Wobbrock, J. O. (2011). In the shadow of misperception: Assistive technology use and social interactions. In *Proceedings of CHI 2011* (pp. 705–714). New York: ACM.

Siebers, T. (2008). *Disability theory*. Ann Arbor: University of Michigan Press.

Siebers, T. (2010). *Disability aesthetics*. Ann Arbor: University of Michigan Press.

Sobchack, V. (2004). A leg to stand on: Prosthetics, metaphor, and materiality. In V. Sobchack, *Carnal thoughts: Embodiment and moving image culture* (pp. 205–225). Berkeley: University of California Press.

Tremain, S. (Ed.). (2005). *Foucault and the government of disability*. Ann Arbor, MI: University of Michigan Press.

Utray, F., de Castro, M., Moreno, L., & Ruiz-Mezcua, B. (2012). Monitoring accessibility services in digital television. *International Journal of Digital Multimedia Broadcasting, 2012*. doi:10.1155/2012/294219

Venkatesh, V., & Brown, S. (2001). A longitudinal investigation of personal computers in homes: Adoption determinants and emerging challenges. *MIS Quarterly, 25*(1), 71–82.

Williamson, B. (2012). Electric moms and quad drivers: People with disabilities buying, making, and using technology in postwar America. *American Studies, 52*(1), 5–29.

Winner, L. (1986). *The whale and the reactor: A search for limits in an age of high technology*. Chicago, IL: University of Chicago Press.

CHAPTER THIRTEEN

The 20th Anniversary of the Digital Divide

Challenges and Opportunities for Communication and the Good Life

SUSAN B. KRETCHMER,[1] PARTNERSHIP FOR PROGRESS
ON THE DIGITAL DIVIDE
JOY PIERCE,[2] UNIVERSITY OF UTAH, USA
LAURA ROBINSON,[3] SANTA CLARA UNIVERSITY, USA

Though born in 1969 at UCLA and Stanford, the Internet did not become commonplace until the 1990s when the 1991 U.S. High Performance Computing Act funded a high-speed fiber optic network that ultimately became the Internet and allowed digital connectivity for computers. This and various technological and business innovations made the Internet easily accessible, navigable, and user-friendly. From 1989 to 1997, the number of households with computers in the U.S. alone jumped from 13.4 to 37.4 million (Kominski & Newburger, 1999), and exploring cyberspace and communicating via email became increasingly useful applications.

At the same time, with the advent of this new technology, officials in the Clinton Administration wondered if there should be concern about equity of access to computers and the "Information Superhighway." As a result, based on Census data about telephone penetration (dial-up telephone lines were then the on-ramp for access to the Internet), computer/modem ownership and usage collected in 1994, the newly created National Telecommunications and Information Administration (NTIA) prepared and released in 1995 the landmark report *Falling Through the Net: A Survey of the "Have Nots" in Rural and Urban America*. Consequently, the inequalities of online access began to be discussed as a new aspect of the larger issues of wealth and poverty, and the "Digital Divide" was recognized as a key challenge in countries around the world.

Twenty years later, rather than leading the charge to connect people to the digital age, the country in which the Internet was invented and that brought the digital divide to global attention instead lags behind 35 other countries, including Korea, Qatar, United Kingdom, Canada, Czech Republic, and Poland, in household Internet access. In addition, the latest available data (ITU, 2014) indicate that in all but 21 out of 228 countries around the world and even in the vast majority of developed countries, at least 20% of the population does not benefit from even minimal access to the Internet at home. Indeed, even within the top 50 countries in Internet access, 29 countries, including Bahrain, New Zealand, Spain, Italy, and Russia, have 20%–38% of their households offline.

That lack of consistent, quality access to emergent communication technologies is antithetical to the nature of a "good life" amidst the transformative changes enjoyed by members of the wired population; those who are offline are alienated from the benefits of the "new opportunities to communicate and interact…new experiences, behaviors, and habits…[and the ability to] engage with others or receive information" suggested by the ICA Conference Theme. Thus, this chapter responds to ICA's call to consider "what a 'good life' might look like in a contemporary, digital, and networked society, and what new challenges we might face in attaining it" to include all members of society. Through new perspectives on the nature of the U.S. digital divide and excerpts from unique sets of case studies, this chapter provides an opportunity to consider the current state and future possibilities to advance the research agenda, inspire new ideas, and identify new areas of necessary, productive focus for the third decade of digital divide research. As well, the exploration of issues in this chapter fosters greater understanding and enlightens practice and policy going forward so that all global citizens can create their own "good life" in the digital, networked age.

THE DIGITAL DIVIDE:
THE PROMISE OF TECHNOLOGY AND THE REALITY PAST, PRESENT, AND FUTURE

Often we assume new technology will revolutionize the world in ways never before imaginable and such was certainly the case with the advent of computers and the Internet. For instance, in 1996, Reed Hundt, then Chairman of the U.S. Federal Communications Commission (FCC), expounded, "In the promise of communications technology we now have a new chance to deliver on the obligation of creating equal opportunity. Networks in poor neighborhoods or rich suburbs equally deliver at lightning speed the learning of the Library of Congress. The dawning of the Information Age represents an opportunity for equality that we have not enjoyed since Horace Mann first championed the idea of the free public school."

Yet, as we now know, as with other technologies before, the promise of information and communication technology did not meet reality. Despite the extensive proliferation of computers and the Internet, widely disparate levels of access and use materialized, and we find that we have only replicated our past in a new form with digital divides as symptoms, reformations, and a new aspect of the larger persistent issues of wealth, poverty, and inequality that have long plagued all countries around the world. Nevertheless, society in general in the 21st century has continually moved toward the digital, often seemingly oblivious to those who are disconnected from the assumed ubiquitous network and, therefore, left behind. As Negroponte (1995, p. 6) observed, "Computing is not about computers any more. It is about living." But, what is the nature of that life with inequity in the digital age and what does it mean for our future?

In 2014, as the Internet marks its 45th year of existence and the World Wide Web its 25th, the digital divide in the United States persists past the 20th anniversary of its recognition through social scientific research and, in myriad ways, stands as a major impediment to the achievement of "the good life" for a large portion of the American public. According to the most recent data, 28% of U.S. households (34.3 million) do not have any Internet access at home (NTIA & ESA, 2013), and 100 million Americans do not have a broadband Internet connection (IMLS, University of Washington, & International City/County Management Association, 2012). While over time the digital divide has narrowed, a very significant gap endures with historical variations across demographics such as race and ethnicity, age, income, education, urban-rural location, and disability status (File, 2013; NTIA & ESA, 2013).

To better understand where we have been and where we are headed, it is instructive to consider the data on the digital divide in a new way and assess the trajectory of efforts to date that have been aimed at creating a more inclusive society. There are several ways to gauge the progress over time of a given group relative to others with respect to Internet access. Table 13.1 utilizes three indicia:

1) the first column shows the percentage-point change (in all cases but one, decrease) in the divide between 2000 and 2011 for the groups compared;
2) the second column shows the decrease in the divide (improvement) from 2000 to 2011 as a percentage of the original 2000 gap (i.e., the expansion rate); and
3) the third column shows the gap in percentage points remaining as of 2011.

As Table 13.1 demonstrates, we have made pitifully little progress in closing the divide over an 11-year period and where there have been more substantial gains, it is a matter of the more advantaged groups being propelled ahead while those who truly need help the most are still left behind.

Table 13.1. Comparison of Digital Divide in Internet Use between 2000 and 2011 by Selected Characteristics.

	Decrease in Divide from 2000 to 2011 (Percentage Points)	Decrease in Divide from 2000 to 2011 as a Percentage of the Original 2000 Gap (i.e., Expansion Rate)	Gap Remaining as of 2011 (Percentage Points)	Number of Future Years Needed at the 2000–2011 Expansion Rate to Close the Gap Remaining in 2011
All Households Without Internet Use Compared to All Households With Internet Use (Percent)	30.2	52%	28.3	6.0
Race and Ethnicity (Percent of Households with Internet Use)				
Black Alone Compared to White Non-Hispanic Alone	3.2	14%	19.3	15.2
Hispanic Compared to White Non-Hispanic Alone	4.6	20%	17.9	9.8
Educational Attainment (Percent of Households with Internet Use)				
Less Than High School Compared to Bachelor's Degree or More	1.3	2%	53.0	294.4
High School Degree (including GEDs) Compared to Bachelor's Degree or More	7.4	20%	28.7	15.8

Some College Compared to Bachelor's Degree or More	4.4	26%	12.6	5.3

Family Income (Percent of Households with Internet Use)

Under $15,000 Compared to $75,000+	12.7	20%	52.0	28.6
$15,000-19,999 Compared to $75,000+	13.9	24%	44.7	20.5
$20,000-24,999 Compared to $75,000+	16.0	30%	38.1	14.0
$25,000-34,999 Compared to $75,000+	12.4	28%	31.7	12.4
$35,000-49,999 Compared to $75,000+	12.3	40%	18.6	5.1
$50,000-74,999 Compared to $75,000+	10.1	60%	6.6	1.2

Metropolitan Status (Percent of Households with Internet Use)

Rural Compared to Urban	0.3	2%	12.0	66.7

Employment Status (Percent of Individuals Age 16+ with Internet Use Anywhere)

Non-Employed Compared to Employed	5.1	19%	21.6	12.5

	Decrease in Divide from 2000 to 2011 (Percentage Points)	Decrease in Divide from 2000 to 2011 as a Percentage of the Original 2000 Gap (i.e., Expansion Rate)	Gap Remaining as of 2011 (Percentage Points)	Number of Future Years Needed at the 2000-2011 Expansion Rate to Close the Gap Remaining in 2011
Disability Status (Percent of Individuals Age 16+ with Internet Use Anywhere)				
Has a Disability Compared to No Disability	Increase in divide of 1	-4%	29.3	Gap will never be closed and may increase

Note. Table created by Susan B. Kretchmer based on data from U.S. Census Bureau, Current Population Survey, August 2000 and July 2011 and U.S. Census Bureau, Survey on Income and Program Participation, research data file (August-November 1999, Wave 11).

Specifically, and most troubling, this analysis reveals that, while the national average expansion rate for household Internet access is 52%, the discrepancy between those with less than a high school education compared to those with a bachelor's degree or more and the gap between rural and urban residents only decreased by 2% from 2000 to 2011 with a 53 and 12 percentage point divide, respectively, remaining in 2011. Worse still, the disparity between those with a disability and those without a disability, now standing at 29.3 percentage points, actually increased by 4% during that same 11-year period. Moreover, we see ascending levels of gain as we go up the ladder of income and education so that those households with $50,000–$74,999 incomes compared to those with $75,000+ incomes closed 60% of the gap in 11 years with only 6.6 percentage points remaining, while in stark contrast, the divide between households with incomes under $15,000 compared to $75,000+ income households improved a mere 20% with a 52 percentage point chasm remaining in 2011. A similarly vast divergence exists when looking at the noted 2% gain made in the gap between those with less than a high school education compared to those with a bachelor's degree or more with a 53 percentage point divide remaining as opposed to the 26% improvement experienced by those with some college compared to those with a bachelor's degree or more where only a 12.6 percentage point difference remains.

Further, as the fourth column of Table 13.1 makes evident, this analysis has significant and serious implications for the future. It teaches us that in real terms, if we continue our current trajectory assuming the initiatives to bridge the divides can maintain their existing rates of success which may not be possible if they reach intractable portions of the gaps, it will take from decades to hundreds of years, if ever, to bring disadvantaged groups into equality in the digital age. For instance, consider that at the present rate it will take 10–16 years for Blacks, Hispanics, those with high school degrees, households with incomes of $20,000–$24,999 and $25,000–$34,999, and the non-employed to reach parity in Internet access with their more advantaged counterparts. But, it will require 2–3 decades for those households with incomes under $15,000 and $15,000–$19,999, almost 67 years for rural residents, and more than 294 years for those with less than a high school education to attain levels of Internet access equivalent to more privileged groups. And, those individuals with disabilities may never overcome the divide and may even find increasing inequality in the future.

ADVANCING THE RESEARCH AGENDA: WHAT WE NEED TO UNDERSTAND

Over the past 20 years since the recognition of the digital divide, in the search for equity in access to "the good life" in the digital age, academic research has played a key role in the public discourse on the issues of the digital divide as well as in the decision-making by policymakers and practitioners as they work to craft solutions to this pressing societal concern. Yet, despite all that we do know so far, there is a very large gap in our understanding of the digital divide, what it means to have and provide access and participation for a diverse population, and how to create a "good life" for all in our digital, networked society. And, that lack of understanding, in turn, impedes our ability to design policy and practice that matches reality, truly addresses needs, and breaks down barriers.

For example, by utilizing regression analysis, we learn that differences in socio-economic and demographic characteristics explain only a small portion of the gaps in adoption between Whites as opposed to Blacks and Hispanics and urban as opposed to rural households (ESA & NTIA, 2011). When controlling for socio-economic and geographic differences, the home broadband adoption gap between Asians and Whites disappears. In contrast, 69% of the gap between White and Black households and 73% of the divide between White and Hispanic households persists even after accounting for demographic, socio-economic, and geographic factors. Further, while, overall, computer users displayed much less disparate home broadband adoption rates across race and ethnicity with a four percentage point gap, controlling for demographic characteristics and geography again erases the Asian and White difference, but both the White-Black and White-Hispanic gaps are still 75% unexplained. Similarly, comparing home broadband adoption between urban and rural households, 39% of the gap for all and 63% of the divide for computer users cannot be accounted for by socio-economic and demographic attributes. Thus, race and ethnicity and urban-rural residence are independently associated with technology usage patterns. But, *why*? What *explains* these divides and the persistent gaps that, as we have seen in the previous section, are so impervious to current policy and practice initiatives and the ever-growing mainstream cultural assumption that computers and the Internet are central to "the good life"?

A full assessment of the gaps in knowledge that need to be addressed (Kretchmer, 2014) reveals how vital research questions, data, and analysis can be overlooked, ignored, or not fully explored even though they have acute impacts on the lives of those struggling to navigate in search of "the good life" in the digital age. These deficiencies cover a broad range, but can be as simple as the absence of versions of surveys in languages other than English and the use of telephone surveys that only reach those who have telephones when the significant percentage of households (10%, possibly more) who do not have consistent telephone access constitute a permanent barrier to universal home broadband (Kretchmer & Schement, 2011)

The balance of this chapter is devoted to exploring three crucial opportunities for research that can improve our understanding of the digital divide:

1) Demographic groups that figure prominently in policy discourse exhibit diverse patterns of access, use, and outcomes with peoples/cultures accessing and constructing distinct media environments in their homes and in their communities. However, policy and practice perspectives that support the value of that diversity have yet to be employed because we do not have a full understanding. This is especially important when considering the diversity of Americans without access, a population only partially described to date and difficult to study by traditional survey methods. One key area for future research, policy, and practice is to assess, understand, and learn from these persistently disconnected groups and the tendencies and tensions inherent in the multi-dimensional tapestry that is the American information age democracy.

2) From the outset, the primary interest and goal of digital divide policy and practice has been universal home Internet access, first as dial-up and now, as technology has evolved, as broadband, because it allows the user as much time and freedom online as s/he desires. Providing access at schools, libraries, and third places is also central to strategy to combat the digital

divide, but those locations are viewed as augments to home access or fall-back for those without home access. We know from statistical analysis that home access correlates, for instance, with higher education and income status. However, we know very little about the difference in the value to the user and outcomes (qualitative, or quantitative (i.e., higher grades, increased reading, etc.) of Internet access at school/library/third places versus at school/library/third places plus at home. Thus, that topic requires further exploration so we can understand the interrelationship of these issues.

3) Because policy discourse tends to focus at the national level—at the center rather than the margins—these diverse patterns are rarely engaged. To remedy this issue, we must turn to not only quantitative research but also qualitative data, such as that contributed by ethnography. While quantitative data cannot capture the context of the scene, that is a major strength of ethnographic research—allowing us to hear the voices and stories of the participants, see behaviors, and get deeper information about how people and groups use technology and create meaning about that use. This level of intensity, richness, and understanding that we would never have been able to achieve otherwise enlightens the social context and cultural reality as well as the shared and discreet identities fostered by a wired world. More-over, harnessing this knowledge helps reframe the underlying assumptions of research, policy, and practice with more equitable value judgments. (Kretchmer & Carveth, 2001)

To explore these issues, we now turn to excerpts from two sites of case studies. The research from the first site provides insights into the tensions and value judgments, everyday struggles, social context, and complexity integral to home, school, and work life for families that are part of a marginalized, disadvantaged population, yet striving to connect to the digital age. The research from the second site illuminates differential uses of the Internet based on the level of access to digital resources and the impact on academic success and intellectual growth demonstrating that at school/library/third places plus at home Internet access is far more valuable to the user than school/library/third places access alone. Both sets of excerpts from larger-scale unique ethnographic studies allow us to see the lived reality of the statistics and research we have explored, to hear directly the voices and perceptions of the participants, and to get a sense of what we have yet to learn. Taken together, this research reveals the importance of filling gaps in our knowledge with a deeper understanding that provides new perspectives on the nature of the digital divide and the complexities, paradigms, challenges, and opportunities of diversity and technology use in the digital age.

VOICES FROM THE MARGINS: SITE ONE
INVISIBLE DIGITAL NATIVES AND THEIR FAMILIES

Prensky (2001, p. 1), echoing Hundt's (1996) promise that technology will create equality for future generations, called school children "digital natives," having spent "their entire lives" engaging with technology through computers, video games, cell phones, and other electronic tools. The experiences described here of working with Mexican immigrant children in a wealthy city indicate "the good life" of being constantly connected, available, tuned in, and benefiting from digital resources as an integral part of home and education are not these children's realities. Does this mean this underrepresented population of digital natives is not living "*The* good life"?

Focusing on an aspect of the digital divide rarely seen, this section features excerpts from case studies that show promises unmet and the complexity of life and sometimes harsh realities poor minority children and their families face while living among mainstream Americans enjoying "the good life" in a networked society. The first excerpt illustrates the tension some families face when having to make serious decisions about needs versus wants. Carmen's story illuminates the real-life everyday struggles that challenge a child when trying to negotiate self-motivation and family expectations. The third excerpt provides a glimpse into Carmen's home life through in-home instruction for her mother Marisol and demonstrates how linguistic and educational limitations, cultural norms, and social expectations are bound up in home, school, and work aspirations for these families. The stories presented are representative of others that reveal the struggles and tensions that co-participants—people in the community, administrators, instructors, and the researcher—confront while navigating computer ownership, instruction, and online exploration.

Data and Methods

The data analyzed here, part of a larger-scale, multi-year ethnographic project launched 15 years ago, are drawn from nearly two years of participant observation in a primarily Mexican immigrant community nestled in a wealthy, technologically connected city—a technopopulous city. The study began by distributing a questionnaire asking about computer usage, ownership, and family demographics. More than 90% of questionnaire respondents did not know how to use a computer and 98% were interested in getting computer instruction through "St. Mary's Ministries" digital literacy program. There were two digital literacy programs provided by the researcher: Computer Club, which took place during St. Mary's afterschool program, and in-home instruction for the children's parents. The second phase of the study employed participant observation and individual interviews with 3rd, 4th, and 5th grade students in the local elementary schools and at least one adult in their households.

Analysis: The Complexity of the Struggle for "The Good Life" in the Digital Age

Wants versus needs: Putting family first. A family that had been given a home computer in the past from St. Mary's needs a new one as they say one of their children spilled something on the computer and the mother threw it away without seeking repair or assistance from St. Mary's. When asked about the possibility of replacement, "Sister Alice," the White, Catholic nun who is Director of Education at St. Mary's, explains, "Oh, that happens sometimes." [Long pause] "It's okay if she doesn't have it anymore. Last year we gave out several computers. Then a mom told us her family's computer was stolen. We felt awful and gave her another one. [Laughing] Then I saw her son in school and he said his father sold it to 'some guy'...I don't know who." Asked if after that St. Mary's took the new computer back, Sister Alice replies, "No! I know that they don't have much and sometimes have to do things that aren't exactly above board. I just wished they would have asked us for help." She goes on to explain that they have a food bank, a clothes bank, and sometimes are able to give small monetary gifts. "That kind of thing doesn't happen often, but I know things are tough for these families. It is hard for some of these families to trust."

Carmen. "Carmen" is absent the first day of the afterschool program because she has to accompany her mother to the doctor. Her classmates explain that it was not Carmen, nor her mother who needed medical care but rather Carmen's sister. Carmen's job was to serve as translator because her mother does not speak English. Despite the rules that each student attend the first session, a spot is retained for Carmen due to her extenuating circumstances. The following week Carmen introduces herself,

> I'm really excited I get to use a computer. [She has a computer at home] but sometimes it's hard for me to use it. My brother and his wife are always on it, but my dad said he is going to get a laptop. Well, maybe a laptop only for him. Will I get to use my own computer in Computer Club?

Upon observation, Carmen learns to navigate web design software quickly despite never having used any software other than course-based keyboarding software required through the school. She easily understands the logic behind the tabs File, Edit, View, and so on. She is one of the first to complete her contribution to the Club's website.

For Carmen, Computer Club is an opportunity to expand her ideas through writing, explore what is available on the Internet, and help organize her classmates so that they can have a "voice on the web" like many of her mainstream, native English-speaking peers.

Marisol. Carmen's family lives in a two-bedroom apartment on the third floor of a multi-unit apartment complex. There is a large, flat screen television against the far wall. A computer table is beside it. Facing the television is a sofa and behind the sofa is a twin-sized bed. A small kitchen table stands between the bed and kitchen counter. A woman and young child are watching television. A man enters the room and takes a seat on a nearby chair. Eight people—four adults and four children—live in the apartment permanently and other relatives and friends reside there temporarily throughout the year.

"Marisol" is Carmen's mother. She does not speak English but understands a few words. A translator conveys that Marisol does not believe she can learn to use a computer because she is illiterate in English and Spanish. The translator explains to Marisol the icons that will assist her in accessing valuable information on the computer and online. The translator asks Marisol what she wants to learn. "Leer...escribir...de todo," Marisol responds. She is excited to learn to read... write...do it all. Her husband sits beside her and acknowledges he too wants to learn more about the computer. He is literate in Spanish; his English-speaking skills are limited as his ability to read and write English is purely functional.

Discussion

Full or even partial participation in digital inclusion as conceived by policymakers and some researchers is not only unrealized for the community of Mexican immigrants in this project, the future for such progress is uncertain. How is always-already changing communication technology useful to children immersed in a rich learning environment with little to no online network connections outside of the classroom, when the community mainstream population makes assumptions without regard to the specific cultural, linguistic, and social needs of that underrepresented population? Through extensive interviews and interaction with children and their families, we begin to see a picture of what one population deems important in life, even if they are unable to articulate how to gain such benefits through technology. While it is possible to focus on the good lives or potential good lives of the invisible digital natives, doing so would whitewash the importance that family and home life plays in what happens in school. Indeed, as the findings of the study excerpted here indicate, lessons learned in the home, community, and through familial relationships across borders are more important than teacher influence and institutional education. Thus, this research suggests we need to focus on ways to motivate teachers and community leaders as well as researchers and local and national policymakers to fully investigate, understand, and support the interests and well-being of invisible digital natives and their families, moving away completely from dichotomies of have and have-not, technological signs and signifiers as a means to stigmatize and impose demands to participate in the digital age. Instead, we must negotiate how to appreciate and celebrate difference while thinking collectively, openly, and practically about how information and communication technologies connect and affect the everyday lives of diverse populations.

VOICES FROM THE MARGINS: SITE TWO
DIGITAL INEQUALITIES AND CAPITALIZING ACTIVITIES

As we have seen earlier in this chapter, economic barriers to Internet access, particularly the means to pay for a computer and home connectivity, are a major component of the digital divide. Economic disadvantage, as both a cause and consequence of digital inequalities, is associated with particular patterns of Internet usage (Witte & Mannon, 2010). Even though there is an abundance of information today, those who lack economic capital often also lack "information capital" (Hamelink, 2001/1999; van Dijk, 2005) and thus find it challenging to operate in the ever-more critical digital realm. At the same time, those without the skills to use the Internet for capital-enhancing activities are more likely to be economically disadvantaged later on—shut out of more remunerative segments of the labor market and, as a consequence, shunted into lower-paying jobs (DiMaggio & Bonikowski, 2008).

Given these digital inequalities and their impacts, it is vital to understand the link between access inequalities, access points, and patterns of usage. Yet, our knowledge of these interrelated issues is sorely lacking. To enrich our understanding, the research excerpted here examines the linkages between digital inequalities and use of the Internet for capital-enhancing activities among high school students setting out on their life journeys. The study shows that, among these youths, use of the Internet for capital-enhancing activities that leads to self-optimization is predicated on favorable conditions of access. In contrast, when access conditions are unfavorable because of economic constraints, students engage with the Internet as information consumers who do not capitalize on digital resources but rather consume them as they would analog resources such as television. This undermines the Internet's potential impact on their lives and has profound ramifications that set the stage for the perpetuation of future disadvantage, putting the "good life" out of reach.

Data and Methods

The data analyzed here are drawn from a larger-scale, multi-year project including 503 ethnographic interviews with students attending two high schools located in agricultural California. One school is a large public institution, a Title 1 or high-poverty school with over half of students eligible for free lunch; the other school does not qualify as a Title 1 school. Students in both schools come from diverse racial and ethnic backgrounds. Gathering data from two schools in this way was key to ensuring that variation is not due to the school setting but rather students' level and site of Internet access. Participants were recruited from a wide range of classes and asked to answer the same battery of questions about their Internet access at home, school, and third places, as well as its use for capital-enhancing activities.

The one-on-one and small-group interviews were grounded in long-term ethnographic fieldwork. To analyze the data, a grounded approach appropriate to emergent conceptualizations and explanatory accounts was employed. The data were coded inductively by moving back and forth between the data and concepts; subsequently, targeted coding was developed, and the data were recoded. Through this process, the patterns of academic Internet use detailed in the analysis section were identified. Using a grounded approach was central to understanding students' use of the Internet from their perspectives (Emerson, 1983). For this reason, ethnographic interviews were well aligned with the study's aim to uncover members' meanings and understandings (Emerson, 2001). A grounded approach also allows the research to preserve members' voices to capture their own lived experiences (Emerson & Pollner, 1988). These processes allowed participants to be classified and compared in terms of their access to information resources and their use of those resources for capital-enhancing activities as capitalizers and non-capitalizers.

Analysis: The Impact of Unequal Levels and Sites of Access to Digital Resources

Capitalizers. Capitalizers parlay access to informational resources into capital-enhancing activities for schoolwork thanks to access to resources. These students enjoy home Internet access with favorable conditions over long periods of time in addition to school/library/third places access, allowing them to approach the Internet as a learning resource and self-teaching tool. As one capitalizer, "Kingsley," underscores, "...I use the Internet mainly for Kahn Academy.com, a site that allows me to go over lessons in math subjects (my weakest subject) and to go over things to study for my SAT/ACT." "Esmeralda" explains, "I have used the Internet for basically all my classes—for English, Math, Chemistry,

Physiology, Anatomy, Economics, Government, and Spanish." "Katia," another student who can access the Internet whenever she wants to pursue capital-enhancing activities, relates, "Every single class I have ever taken through my high school education wasn't necessarily Internet dependent, but I did benefit greatly from having Internet access which was a crucial step towards thoroughly understanding the course..."

Even more important, capitalizers use the Internet in sophisticated ways that result in self-directed learning. For example, "Jordan" enhances complex thinking skills by using the Internet as a sophisticated tool to compare two poems for her English class,

> I looked up the poems first and the symbolism and by doing this I was able to come to my own conclusion of what the poems were portraying...If I did not have a computer to research this, my essay would lack in substance...I use the Internet to research the topic and possibly give me new perspectives and ideas that I had not thought of. The assistance that the Internet gives me is great because it really is a learning tool. Whether or not the information I come across correlates to the assignment given, I am able to still educate myself and even clarify things that I had questions on.

As this indicates, whether or not they are technically required to use the Internet for schoolwork, these students assume that Internet use is an integral part of every academic class. "Jesús" explains,

> When I use it [Internet] for physics homework...it helps reinforce what I already knew and helps me understand the process of getting to an answer or understand other issues involving the problem. Also helps me get ahead...So, I'm gonna use it [Internet] pretty much no matter what...the coursework is rigorous and often requires a vast amount of research to be completed for any given assignment. The access to Sparknotes and other study guide websites is extremely helpful and eases the workload a bit more...

For such students, the Internet functions as far more than a tool to convey or retrieve predefined information. They deploy the Internet to generate ideas, to compare and adjudicate between information, and to seek additional knowledge with which to educate themselves about topics beyond the scope of immediate tasks or assignments at hand. In so doing, Internet resources serve as the ultimate self-optimization tool for intellectual growth.

Non-Capitalizers. The Internet does not play the same optimizing role for those who lack favorable access conditions. Non-capitalizing students lack home Internet access via any digital device. Obtaining even basic access to the Internet for these students necessitates going to school computer labs, the public library, or third places. "Lloyd" expresses his frustration at not having home Internet access, "It is harder in my situation because I have never had the opportunity to have these luxuries; it is very difficult for me to get hold of a computer with Internet..."

"Morris" tells a similar story, "I think I could use the Internet for my English homework...like other kids...but I gotta go to work after school so I can't go to the lab or the library..." Non-capitalizers explain that occasionally teachers have taken them to the school computer lab. Indicating the paucity of such occasions, "Pat" recalls, "Yeah, we went to the [school computer] lab two times for my English class last term but that was it." In the words of one digitally deprived student, "I want to use the Internet more but I don't have it at home...so I don't really get to use it for school much."

Significantly, constant resource scarcities keep non-capitalizers' Internet experiences limited to the realm of Web 1.0. Because of their limited access, these students take a more passive approach to information gathering on the Internet. Indeed, even when they do have Internet access opportunities at school, their narratives of Internet use resemble narratives about accessing pre-digital media such as magazines. When Lloyd was asked to do research for his school assignment, he only went so far as to gather images. As he recounts,

> The assignment was to do a presentation on countries that you would like to travel to in *The Amazing Race*. I researched info on the countries and downloaded pictures that were relevant to the countries I went to for the project. The computer and Internet were really helpful.

"Janice" relates a similar experience in which she used the Internet as if it were a book,

> I had to search up a recipe for my cooking class. The Internet helped me find good recipes...pictures for my cookbook...last year I had to research for my government class...the Internet was really helpful because it gave me correct instant information.

Pat treats her Internet forays as experiences in passive entertainment similar to the experience of watching television. When her economics class used the computers in the school library, "It was like way better than readin' a book. It was more like watchin' TV...just sit back and enjoy the ride." Lacking the opportunities to engage with the Internet for non-task related activities as an exercise in self-teaching and self-capitalization, these students' Internet use mimics consumption of other non-interactive media. Under these circumstances, their Internet use is circumscribed in ways that will harm their academic performance, intellectual growth, and self-development.

Discussion

As these findings demonstrate, students' use of the Internet for capital-enhancing activities cannot be understood apart from their access situations. Students with home access to Internet resources in addition to school/library/third places access engage with digital media in a fundamentally different way than do students without access at home. Those with home access go far beyond mere passive information retrieval. Rather, they are able to use their Internet engagements for self-teaching in a very active way. By contrast, disadvantaged students without the time and freedom afforded by home access treat the Internet as they would analog channels that provide information to be consumed. For them, the Internet loses its unique strengths as an interactive medium that allows students to enlarge their horizons and develop their intellectual capacities. In this way, this research indicates how first-order digital divides related to levels and sites of access contribute to second-order gaps related to capital-enhancing activities. Here we see how socio-economic disparities foster differential Internet access that can lead to yet other inequities down the line as offline and online inequalities are mutually reinforcing and self-perpetuating (Robinson, 2014).

TOWARD "THE GOOD LIFE" FOR ALL IN THE DIGITAL AGE

The good life is a process, not a state of being.
It is a direction not a destination.

—CARL R. ROGERS (1995, P. 186)

In this chapter, we explored the promise of information and communication technology and the reality and magnitude of the enduring digital divide past, present, and future by assessing the data on digital disparities in a new way that allowed us to recognize the problematic trajectory of efforts to date that have been aimed at creating a more inclusive society. We also considered gaps in our current understanding of the issues and how we might advance the research agenda by inspiring new ideas and identifying new areas of necessary, productive focus for digital divide research. Finally, highlighting three crucial opportunities for research that can improve our understanding, two sets of excerpts from unique ethnographic studies enlightened the quantitative/statistical data and brought them to life through authentic and rich narratives. These studies provide new insights into what the digital divide and computer and Internet technology use *mean* for the people affected by it. And, they broaden and deepen our understanding of diversity in the digital age through the eyes, voices, lived experience, and perspective of those facing the challenges and trying to gain the opportunities of the digital age. We saw the stark division between those who are able to benefit from Internet resources and those who are not, but also the depth, complexities, nuances, and myriad multi-faceted, multi-layered, and inextricably intertwined issues at that nexus.

The Obama administration has used the language of a social contract to advance their commitment to rebuilding America's infrastructure and extending that vision to the broadband Internet network. The implementation of that vision, however, must go beyond simple connectivity to embrace the public interest goal of ensuring diversity through a promise of common welfare that visualizes information networks as serving everyone and corresponding to the purposes of a just society obligated to create "the good life" for all. But, as the ICA Conference Theme asks, what should "the good life" in contemporary, digital, networked society look like and how do we attain it?

As Carl Rogers suggests, "the good life" is not a fixed state, or some sort of destination of contentment, nirvana, or happiness at which people arrive and stay. Rather, as this chapter makes clear, in the digital age, individuals and communities need to have the tools and conditions that will allow them to strive for and reach their own unique "good life" as they envision it. We might concentrate on reconciling the lived reality of disadvantaged, marginalized populations and the mainstream's assumptions about them made without regard to the specific cultural, linguistic, and social needs of those underrepresented groups so that we can provide the tools and conditions to expand the concept of "the good life" to include a process that honors diversity and enables and values varied permutations of "the good life." Or, we might think about this in the context of the powerful need for and opportunities gained through home computer and Internet technology and the devastating impact of constraints on or lack of access to those digital resources for students and, thus, providing the tools and conditions for more equitable access so that students can find the direction to "the good life." And, as teachers, students, and scholars, we might rise to the challenge of the persistent digital divide by investigating the gaps in our knowledge and what it means to have and provide access, opportunities, and equality for diverse, underserved populations to foster greater understanding, inform policymaking and practice, and find solutions to this pressing societal concern so that all global citizens can have the tools and conditions, the process and direction, they need to create their own "good life" in the digital, networked age.

NOTES

1. A portion of this work draws upon Kretchmer, S. B. (2015). Foreword. In J. Pierce, *Digital fusion: A society beyond blind inclusion*. New York: Peter Lang.
2. Pierce and Robinson contributed equally to this chapter and, therefore, their names appear alphabetically in the authorship listing. Pierce authored the "Voices From the Margins: Site One" section of this chapter, which draws upon Pierce, J. (2015). *Digital fusion: A society beyond blind inclusion*. New York: Peter Lang.
3. Pierce and Robinson contributed equally to this chapter and, therefore, their names appear alphabetically in the authorship listing. Robinson authored the "Voices From the Margins: Site Two" section of this chapter.

REFERENCES

DiMaggio, P., & Bonikowski, B. (2008). Make money surfing the Web? The impact of Internet use on the earnings of U.S. workers. *American Sociological Review, 73*(2), 227–250.

Economics and Statistics Administration (ESA), & National Telecommunications and Information Administration (NTIA). (2011, November). *Exploring the digital nation: Computer and internet use at home.* Washington, DC: U.S. Department of Commerce. Retrieved from http://www.ntia.doc.gov/files/ntia/publications/exploring_the_digital_nation_computer_and_internet_use_at_home_11092011.pdf

Emerson, R. (1983). *Contemporary field research: A collection of readings.* Prospect Heights, IL: Waveland Press.

Emerson, R. (2001). *Contemporary field research: Perspectives and formulations.* Prospect Heights, IL: Waveland Press.

Emerson, R. M., & Pollner, M. (1988). On the uses of members' responses to researchers' accounts. *Human Organization, 47*(3), 189–198.

File, T. (2013, May). *Computer and internet use in the United States.* Current Population Survey Reports, P20–568. Washington, DC: U.S. Census Bureau, U.S. Department of Commerce, Economics and Statistics Administration. Retrieved from http://www.census.gov/prod/2013pubs/p20-569.pdf

Hamelink, C. J. (2001). *The ethics of cyberspace.* (C. J. Hamelink, Trans.). London: Sage. (Original work published 1999)

Hundt, R. E. (1996, May 2). *The school bell merger: Communications and education.* Speech at the Seventh Annual Milken Family Foundation National Education Conference, Los Angeles, California. Retrieved from http://transition.fcc.gov/Speeches/Hundt/spreh623.txt

Institute of Museum and Library Services (IMLS), University of Washington, & International City/County Management Association. (2012, January). *Building digital communities: A framework for action.* Washington, DC: Institute of Museum and Library Services. Retrieved from http://www.imls.gov/assets/1/workflow_staging/AssetManager/2140.PDF

International Telecommunication Union (ITU). (2014, June). *World telecommunication/ICT indicators database 2014* (18th ed.). Available from http://www.itu.int/en/ITU-D/Statistics/Pages/publications/wtid.aspx

Kominski, R., & Newburger, E. (1999, August). *Access denied: Changes in computer ownership and use: 1984–1997.* Paper presented at the American Sociological Association Annual, Chicago, IL. Retrieved from http://www.census.gov/population/socdemo/computer/confpap99.pdf

Kretchmer, S. B. (2014, May). *Is the persistent digital divide a permanent barrier to the "good life"?: Exploring two decades of research, policy, and practice.* Paper presented at the International Communication Association Annual Conference, Seattle, WA.

Kretchmer, S. B., & Carveth, R. (2001). "The color of the Net: African Americans, race and cyberspace." *Computers and Society,* September, 9–14.

Kretchmer, S. B., & Schement, J. R. (2011, October). *Understanding divides, access, participation, and diversity in the digital age.* Paper presented at the Association of Internet Researchers Annual Conference, Seattle, WA.

National Telecommunications and Information Administration (NTIA), & Economics and Statistics Administration (ESA). (2013, June). *Exploring the digital nation: America's emerging online experience.* Washington, DC: U.S. Department of Commerce. Retrieved from http://www.ntia.

doc.gov/files/ntia/publications/exploring_the_digital_nation_-_americas_emerging_online_experience.pdf

Negroponte, N. (1995). *Being digital*. New York: Knopf.

Prensky, M. (2001). Digital natives, digital immigrants. *On the Horizon, 9*(5), 1–6.

Robinson, L. (2014). Freeways, detours and dead ends: Search journeys among disadvantaged youth. *New Media & Society, 16*(2), 234–251.

Rogers, C. R. (1995). *On becoming a person: A therapist's view of psychotherapy*. New York: Mariner.

U.S. Census Bureau. (1999, August-November). 1996 Survey on income and program participation (Wave 11) [Data file]. Washington, DC: The Bureau.

U.S. Census Bureau. (2000, August). Current population survey [Data file]. Washington, DC: The Bureau.

U.S. Census Bureau. (2011, July). Current population survey [Data file]. Washington, DC: The Bureau.

van Dijk, J. A. G. M. (2005). *The deepening divide: Inequality in the information society*. Thousand Oaks, CA: Sage.

Witte, J. C., & Mannon, S. E. (2010). *The Internet and social inequalities*. New York: Routledge.

CHAPTER FOURTEEN

Liberating Structures

Engaging Everyone to Build a Good Life Together

HENRI LIPMANOWICZ, LIBERATING STRUCTURES PRESS, USA
ARVIND SINGHAL, UNIVERSITY OF TEXAS AT EL PASO, USA
KEITH MCCANDLESS, LIBERATING STRUCTURES PRESS, USA
HUA WANG, UNIVERSITY AT BUFFALO,
THE STATE UNIVERSITY OF NEW YORK, USA

"...The world is changed through small, elegant shifts in the protocols of how we meet, plan, conference, and relate to each other. The genius of this [Liberating Structures approach]…is how it puts in the hands of every leader and every citizen the facilitative power that was once reserved for the trained expert." Peter Block on liberating structures (Lipmanowicz & McCandless, 2014, back cover).

Have you been to classrooms with rows of tables and chairs neatly arranged, the students sitting there with their fingers glued to the smartphone while "the sage on the stage" is lecturing away—a lot of bodies that are present but minds that may be absent? Have you been to meetings where discussions are managed by the chair and the entire group spends the whole time listening to just one person talking— perhaps too much is said yet too little is accomplished? These are challenges that we, as communication professors, researchers, and practitioners, face routinely in our professional lives. In this chapter, we discuss the limitations of traditional group communication methods and present Liberating Structures as an alternative or complementary approach to unleash the potential of everyone, increase work efficiency and productivity, and build trusting and generative relationships—with emergent processes, liberating experiences, surprising outcomes, and meaningful connections—one way to build a good life together!

When it comes to the conduct of meetings, whether in classrooms or boardrooms, five methods are commonly used to organize how groups of people work together: (1) the ubiquitous *presentation* with one person in control of the microphone—often the invited expert or the "shower and teller;" (2) the go-around *status report* with the microphone being passed from one person to another (i.e., turn-taking) with the purpose of briefing the boss or the bigger group; (3) the *managed discussion* with one person in charge of coordinating the conversation—often used for consensus-building or decision-making; (4) the *open discussion* with no one in charge but often in response to a presentation or a non-directed question; and (5) the free-flowing *brainstorming*, generating wild ideas through a Ping-Pong style conversation that is too loosely structured and that often misses multiple perspectives or the local know-how (Lipmanowicz & McCandless, 2014).

These five dominant methods of organizing group work severely limit what groups are able to accomplish. They direct the flow of expertise from the top, foster passive acceptance by restricting and controlling participation, and make exclusion a routine fixture of the classroom or any workplace. As a result, group work is deeply frustrating, marginalizing, and oppressive. This is one reason why most of us hate meetings, considering them as a waste of time, resources, and energy. How can classrooms and workplaces become places where people feel engaged? Here, we describe Liberating Structures that make it possible to include and engage all who are affected in shaping their next steps.

WHAT ARE LIBERATING STRUCTURES?

Liberating Structures are simple protocols that groups can use to organize how they work or learn together. Each protocol specifies five structural elements: (1) The *structuring invitation* such as a question to create a common focus; (2) *Space arrangement*, usually an open physical setting is required; (3) *Participation distribution* to ensure everyone has an equal chance to contribute, (4) *Groups' configuration* with different group sizes for different purposes, and (5) the *sequence of steps* and *time allocation* for effective execution. Currently, there are three dozen Liberating Structures available (http://liberatingstructures.com). They are simpler than a process and more serious than a fun exercise. They facilitate the minimum specifications for a group to make progress together without a predetermined outcome. They control the form or structure of micro-interactions in a way that liberates simultaneous mutual shaping of insights and next steps.

A flock of geese flying in a V-formation can illustrate what Liberating Structures make possible to enhance the performance of any group (see Figure 14.1). Whereas a single goose is exhausted after flying 500 miles, a flock of geese flying in a V-formation can fly from 800 to 1,000 miles without resting.

Figure 14.1. A flock of geese flying in a V-formation.

What makes this possible?

Simply, the geese flying in the back utilize the air currents coming from the wings of the geese in front to lift themselves (Papa, Singhal, & Papa, 2006). The geese rotate leadership at regular intervals. When the leader goose tires, it routinely drops behind in the formation as the geese at the back sequentially move forward. This means that if a goose moves out of formation, the increased drag on its wings provides instant feedback to self-correct its position. When in flight, the geese honk regularly and loudly to identify their respective positions and to encourage others to keep going, especially the leader. If a goose is wounded or unwell, two or three geese accompany it to the ground. Once nourished back to health, they will join another passing flock.

So a flock of flying geese maximizes both individual well-being and overall group performance. In the parlance of industrial engineering, a flock of flying geese represents a dynamic, interactive, and collaborative model of ergonomic design, a scientific discipline concerned with the understanding of interactions among actors and other elements of a system in order to optimize the performance of each individual and the overall system. At any given time, each goose has a specified role and responsibility, but across the spread of time, roles and responsibilities, including leadership, are constantly rotating. Effort, participation, and contributions are distributed and balanced across time and distance. There is no wasted effort. All geese are engaged at all times, working in parallel toward a shared purpose. Feedback is plentiful, authentic, immediate, and affirmative. The geese are ever mindful of not just *who* they are, but *whose* they are!

WHY IS WORK LIFE OFTEN "BAD LIFE"? HOW LIBERATING STRUCTURES CAN CREATE "GOOD LIFE"!

Akin to the rotating V-formation of a flock of flying geese, Liberating Structures specify how each participant's time, effort, and contribution are distributed in different spatial configurations so that everyone has an equal opportunity to participate, dialogue, and shape the group decisions and outcomes. However, the standard and dominant practice in a classroom or workplace is a far cry from what is embodied in a flock of flying geese, or embedded in the premise of Liberating Structures.

The designs of classrooms, boardrooms, and workspaces are deeply rooted in the ideology of the Industrial Revolution, emphasizing standardization, uniformity, and regularity. Participants, sitting in rows and columns, should behave in an orderly manner. Students and employees are looked upon as commodities to be processed, trained, programmed, and produced in an invariant manner. This widespread notion that students and employees are throughput and commodities needs to be challenged. Liberating Structures challenge the prevailing notion that a workplace cannot be engaging or enjoyable. In fact, when participants are engaged in a work place, productivity and group performance outcomes are significantly higher (see Figure 14.2).

Figure 14.2. Multiple small circles in Singhal's class in Tokyo, Japan in 2011.

If group performance can be significantly enhanced, and work be made more enjoyable, why hasn't it happened much? Here are some clues, based on our collective experience in educational, corporate, and non-profit settings:

First, routine work practices are so normalized that they are pretty much invisible. They are what everybody does. They are not diagnosed as a big source of problems or opportunity. They are not on anyone's radar screen. If you have never seen high engagement, how can you believe it even exists or is possible? If you are not convinced it exists, why would you bother looking for it or looking for a way to create it?

Second, improving the level of engagement in an organization is perceived as a big complex challenge. The dominant thinking is that it requires big complex programs, culture change campaigns, extensive leadership development, possibly reorganizations, or a new cadre of leaders. Small chicken-shit changes in routine practices are totally absent from the slate of solutions.

Third, we are all simply doing what we know how to do. We are doing mostly the same thing as the people above/before us are doing. In the hierarchical model within which we all grew up, people at the top are telling others what to do. They are expected to know all the right answers (experts) and to be competent at directing others (parenting, educating, inspiring, managing, leading). We all know that reality is different, but in the absence of something else, we continue to perpetuate the same organizational model for school, work, home, and church, etc. This model is not inclusive, it includes a lot of "shut up and listen."

Fourth, inertia is enormous for the very reason that the current standard practices are totally imbedded in the daily functioning of nearly all organizations, from top to bottom and across all functions. To appreciate the weight of inertia, it is enough to look at boardrooms where elongated tables occupy most of the space, sitting arrangements are cast in stone, and all meetings look the same, exact same structure, just different agenda items. That is the model that cascades down into organizations of all kinds.

Last but not least is fear of the consequences of doing something different. The existence of practices such as Liberating Structures is not widely known. The first book about them (Lipmanowicz & McCandless, 2014) was recently published. Liberating Structures usually are a visible departure from the prevailing habits, traditions, and culture. For new users, they can easily be a source of anxiety: Until others around me see their benefits, how will they react? What if "it" doesn't work? What will people think of me? What do I do if people get confused or refuse to participate?

Our experience suggests that Liberating Structures not only ensure that people who are more positive and creative will get the space they need to contribute but they also invite the better side of all participants to show up. When people experience new patterns of interactions and see the results, it invites them to experiment with new practices. When their voices are heard, participants feel valued, and are motivated to do more. In short, Liberating Structures create the conditions for a healthier ecosystem to emerge.

THE CONCEPTUAL BASIS OF LIBERATING STRUCTURES

The conceptual basis of Liberating Structures can be traced back to the teachings of the noted Greek philosopher Socrates over two thousand years ago and more recently to noted 20th century educational practitioners and scholars such as Dewey (1987/1938), Bruner (1960, 1973, 1996), Piaget (2001/1947), and Montessori (1986). All of them, in their own way, critiqued the industrial model of public education that privileged expert knowledge and overly emphasized delivery of content rather than paying attention to process, experience, and self-discovery (Kolb, 1984). They all deeply valued hands-on, experiential discovery, emphasizing the importance of interactions, dialogue, and collaboration in the learning process. Principally, they argued for curriculum to be organized in an upwardly spiraling manner so that the student continually builds upon what they have already learned (Darling-Hammond, 2013; Davis, 2013). They emphasized effective sequences in which to present material so that the learning emerged from the students' own curiosity-fueled engagement, not from invariant transmission of expert knowledge.

Despite the valorizing of principles espoused by Dewey, Bruner, Piaget, Montessori, and others, our educational institutions treat students as empty vessels to be filled with the expert knowledge (Freire, 1971/1968). In workplaces, usually it is the superiors who speak and direct; subordinates listen and comply. Unwittingly, conventional structures stifle inclusion and engagement. Meetings and group work lead to disengaged participants, dysfunctional groups, and wasted ideas. Liberating Structures allow participants to recover their voices and agency and help them *discover and* believe they have something worth saying.

TRANSFORMING CLASSROOMS AND LEARNING EXPERIENCES

A small example illustrates concretely what happens when someone uses a Liberating Structure. Anu is a teaching assistant giving a course in public speaking to a group of some 30 undergraduate students at a medium-sized Southwestern university. After all students had their first public speaking experience she wants to do a quick debrief, have the students reflect on what they learned and, looking back, what they would do differently. A standard practice, the one that has been in use for centuries, would be for the teacher to throw the question at the whole class.

A few students would raise their hands. She would pick three, four, or perhaps five students to share their thoughts, and then she would share her own observations and recommendations. All other students would be left with no choice but to listen passively. Most students would have likely zoned out.

Anu instead decided to use a Liberating Structure called Impromptu Networking (http://www.liberatingstructures.com/2-impromptu-networking/). She first asked each student in class to stand up. Then she told them that she wanted them to pair up preferably with someone they didn't know well and that they had 30 seconds each to answer the following question, "Looking back at your first public speaking experience what would you do differently?" She told the students that after the first round she would ring a bell and they would have to pair up with another student for another 30 seconds while addressing the same question. And then there would be a third round. Then she rang her bell and said, "Go, first round!" The whole room erupted in spontaneous combustion. All students were immediately engaged first sharing their idea and then listening to their partner.

The energy in the room was palpable. Positive body language was everywhere: students leaning in, smiling, and listening. Three times meant three opportunities to reflect more deeply and learn from peers. At the end of the three minutes, while students were still standing up, Anu asked, "Who would like to share something you heard from a partner that you thought was particularly valuable?" She let the sharing go till it ended on its own; all the learning from the whole class was captured effortlessly and quickly within a couple of minutes. Importantly, what Anu did with 30 students could have been done with 60 or 300 students within more or less the same amount of time. Liberating Structures scale very easily.

This small example illustrates how and why it is possible to be more effective and productive as a group and, at the same time, make it also enjoyable for all participants. It is enjoyable because everybody is actively engaged from start to finish. It feels good because everybody is given equal space to speak and be heard. It is fun because it is dynamic and energizing. It is rewarding because it gives everybody the opportunity to contribute to the whole learning process. It generates lots of interactions between people who otherwise would remain distant in spite of sitting in the same room. These multiple interactions build connections and, gradually, trust between people thus fostering a sense of community, something to look forward to spending time with. Allowing the entire variety of contributions to emerge from the group enriches the conversations while leveling the playing field. The teacher becomes more of a facilitator, a partner in discovering solutions, a co-conspirator in how to have a good time while working together.

Ask any student of Anu's class whether they enjoyed their Impromptu interactions and you will find out why her class is a favorite of theirs, one that they hate to miss. You will also understand why Anu received the university-wide outstanding teaching assistant award for the 2013–2014 academic year. One important twist: Anu had never taken a course in public speaking or practiced public speaking. She didn't teach from a position of expertise. Instead she created the conditions and facilitated the learning of the students by getting them all engaged with Liberating Structures.

While Impromptu Networking is one of the simpler Liberating Structures, it is illustrative of the whole set. A small, discrete example like Impromptu Networking makes it easy to see the differences between a standard instructional practice in a classroom and a Liberating Structure. Those differences remain the same at a larger scale, in more complex situations, and when using multiple Liberating Structures. The differences scale because these engagement outcomes are automatic "side-effects" of the way all Liberating Structures are constructed: get everybody engaged from start to finish, give everybody equal space to be heard and contribute and practice self-discovery. Just as low participation is built into the fabric of standard work practices, high engagement is built into the fabric of Liberating Structures. Table 14.1 lists some of the other commonly-used liberating structures in classrooms.

Table 14.1. Commonly Used Liberating Structures in Classrooms.

Brief Description	Icon	Example of Classroom Use
1-2-4-All Engage everyone simultaneously in generating questions, ideas, and suggestions.	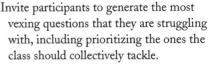	Invite participants to generate the most vexing questions that they are struggling with, including prioritizing the ones the class should collectively tackle.
Conversation Café Engage everyone in making sense of profound challenges.		Invite participants to discuss how to tackle their most challenging questions by expanding and deepening the solution space.
User Experience Fishbowl Share know-how gained from experience with a larger community.		Invite groups to share their unique field experiences, insights, and struggles with the whole class.
Troika Consulting Get practical and imaginative help from two colleagues immediately.		Obtain help on an individual project, assignment, or task from peers, and in turn serve as a consultant to address their challenges.
25-10 Crowd Sourcing Rapidly generate and sift a group's most powerful and actionable ideas.	25/10	Invite participants to rapidly generate the most concrete scenarios to go from knowledge-to-action.

What, So What, Now What?
W³ Debrief together, look back on progress to date and decide actionable next steps.

Analyze a case study in class by step-wise, beginning with a discussion of (1) *what* happened (i.e. establish the facts), (2) *so what* (i.e. discuss inferences and conclusions), and (3) now what (i.e. chart implications for applying the findings). Or, simply, use to track class progress with respect to a particular topic.

Note: More description of each of these liberating structures, including how and when to use, can be found in (Lipmanowicz & McCandless, 2014) and at www.liberatingstructures.com.

Two of the present co-authors, Singhal and Wang, have under the guidance and mentoring of the other co-authors, Lipmanowicz and McCandless, been employing the practice of Liberating Structures to liberate their classrooms—whether in El Paso or Buffalo, or in other parts of the world. Here co-author Singhal, in first person, shares how he makes small adjustments in the protocols of how his classroom is structured and conducted (for more, see Singhal, 2014).

> In my classes, participants invariably find themselves in circles (see Figure 14.2). There is no "sage on stage," and all participants have equal opportunity to be seen and heard. To deepen classroom conversations, I often introduce a "talking stick" when doing small-group work. Whoever holds the stick talks, the others listen. The stick is then passed until all have spoken. The stick may go around three to four times so that participants have an opportunity to widen and deepen their own thoughts and to build upon others' thoughts. Trust rises as relationships deepen over time.

INVITING EVERYONE TO SAVE LIVES

Liberating takes courage. The first wave of mutually shaped insights, decisions, actions, and agreements may seem inconspicuous, crude, or fleeting. They often come from overlooked details, unusual suspects, and need to be coaxed out of messy or ambiguous situations. It is much easier to see big system failures (and then propose standardized outcomes) than to notice how widely distributed local solutions make a difference. System problems shout, widely distributed solutions whisper.

For decades, a standardized approach to preventing the spread of antibiotic resistance organisms (aka superbugs) was delivering modest to poor results. Scientific evidence supported three effective prevention strategies: hand washing, cleaning surfaces, and isolating patients with infections. Standardized policies and procedures regarding *what to do* were developed and handed down from technical experts to the staff interacting with patients and families every day.

With these outcomes predetermined, training to reduce variation in *what to do* was handed down the chain of command. The goal was to tightly manage execution—rewarding adherence and punishing non-compliance. If results were poor, managers and technical experts employed more training, more rewards, and more punishments. If performance still did not improve, more forceful edicts and still more technical *what to do* education was repeated ad infinitum. An unproductive self-reinforcing pattern of over-control or over-helping from and dependency from the front line can take hold.

In contrast, an action research project using Liberating Structures such as Improv Prototyping made it possible for the managers and experts to include the people closest to the challenge in shaping *how to* prevent the spread of infections together (Singhal, Buscell, & Lindberg, 2014; Singhal, McCandless, Buscell, & Lindberg, 2009). For the first time, unusual suspects like cleaners, aides, doctors, patients, and family members were asked: How they knew the risk of transmission was present; what they did to prevent transmission (e.g., how they washed their hands before and after every exposure to patients or unprotected surfaces); what made it difficult to take precautions all the time, and what more they could do to improve or invent new solutions.

Answers, ideas, and small solutions poured out. Many people were astonished that they were being asked. Rarely if ever had they been invited to shape next steps. Being told what to do was far more familiar. New connections within and across functions started to generate results. With more freedom, people were taking more responsibility for solving the problem and working in partnership across barriers.

Paradoxically, the scientific evidence or evidenced-based-medicine about *what to do* was present but the *how to* generate local practice-based-evidence was sorely neglected. The traditions of waiting for direction from the top, power differences among staff, and diverse functional roles created barriers to generating solutions. However, social skills required to work productively with these challenges were acquired rapidly through use of Liberating Structures.

At the start, it was messy and ambiguous. Managers and experts did not know how to ask for help. The cleaners, aides, and patients were not sure their contributions would be valued. Differences in power, social background, and perspectives were enormous. As local action was undertaken, social proof that the approach was working spread quickly. If one unit was able to see their ideas enacted and they reduced transmissions, their more liberated partnership quickly spread to other units. A big problem was being solved and a new way of solving problems together was discovered. Does "the good life" get any better than that?!

IN CONCLUSION

Through our collective work over the past decade or so, more than thirty Liberating Structures are documented. They are precisely described from their particular range of purposes to the details of how to use them. Liberating Structures can be used singly in routine situations. For more sizeable projects or ambitious goals, they can be combined into an infinite range of combinations or strings.

Our experience also suggests that the use of Liberating Structures doesn't demand any exceptional qualities or leadership talents. The structures are so simple that anybody at any level can do it. They don't require extensive training. Liberating Structures don't ask of leaders to develop new and complex competencies. They ask of people to do something that they can do, namely to modify in small ways the practices they use routinely when working together. See Table 14.2 for the ten principles of Liberating Structures.

Table 14.2. Ten Principles of Liberating Structures.

When Liberating Structures are part of everyday interactions, it is possible to:	Liberating Structures make it possible to: START or AMPLIFY these practices that address opportunities and challenges with much more input and support:	Liberating Structures make it possible to: STOP or REDUCE these "autopilot" practices that are encouraged by conventional microstructures:
1. Include and Unleash Everyone	Invite everyone touched by a challenge to share possible solutions or invent new approaches together. Actively reach across silos and levels, beyond the usual suspects.	Separate deciders from doers. Appoint a few to design an "elegant solution" and then tell all others to implement it after the fact. Force buy-in. Confront resistance with hours of PowerPoint presentations.
2. Practice Deep Respect for People and Local Solutions	Engage the people *doing the work* and familiar with the local context. Trust and unleash their collective expertise and inventiveness to solve complex challenges. Let go of the compulsion to control.	Import *best practices*, drive *buy-in*, or assume people need more training. Value experts and computer systems over local people and know-how.

3. Never Start Without a Clear Purpose	Dig deep for what is important and meaningful to you and to others. Use *Nine Whys* routinely. Take time to include everyone in crafting an unambiguous statement of the deepest need for your work.	Maintain ambiguity by using jargon. Substitute a safe short-term goal or cautious means-to-an-end statement for a deep need or a bold reason to exist. Impose your purpose on others.
4. Build Trust as You Go	Cultivate a trusting group climate where speaking the truth is valued and shared ownership is the goal. Sift ideas and make decisions using input from everyone. Practice "nothing about me without me." Be a leader and a follower.	Over-help or overcontrol the work of others. Respond to ideas from others with cynicism, ridicule, criticism, or punishment. Praise and then just pretend to follow the ideas of others.
5. Learn by Failing Forward	Debrief every step. Make it safe to speak up. Discover positive variation. Include and unleash everyone as you innovate, including clients, customers, and suppliers. Take risks safely.	Focus on doing and deciding. Avoid difficult conversations and gloss over failures. Punish risk-takers when unknowable surprises pop up.
6. Practice Self-Discovery Within a Group	Engage groups to the maximum degree in discovering solutions on their own. Increase diversity to spur creativity, broaden potential solutions, and enrich peer-to-peer learning. Encourage experiments on multiple tracks.	Impose solutions from the top. Let experts "educate" and tell people what to do. Assume that people resist change no matter what. Substitute laminated signs for conversation. Exclude frontline people from innovating and problem solving.
7. Amplify Freedom AND Responsibility	Specify minimum constraints and let go of overcontrol. Use the power of invitation. Value fast experiments over playing it safe. Track progress rigorously and feed back results to all. Expose and celebrate mistakes as sources of progress.	Allow people to work without structure, such as a clear purpose or minimum specifications. Let rules and procedures stifle initiative. Ignore the value of people's understanding how their work affects one another. Keep frontline staff in the dark about performance data.

8. Emphasize Possibilities: Believe Before You See	Expose what is working well. Focus on what can be accomplished now with the imagination and materials at hand. Take the next steps that lead to creativity and renewal.	Focus on what's wrong. Wait for all the barriers to come down or for ideal conditions to emerge. Work on changing *the whole system* all at once.
9. Invite Creative Destruction to Enable Innovation	Convene conversations about what is keeping people from working on the essence of their work. Remove the barriers even when it feels like heresy. Make it easy for people to deal with their fears.	Avoid or delay stopping the behaviors, practices, and policies that are revealed as barriers. Assume obstacles don't matter or can't be removed.
10. Engage in Seriously Playful Curiosity	Stir things up—with levity, paradoxical questions, and improv—to spark a deep exploration of current practices and latent innovations. Make working together both demanding and inviting.	Keep it simple by deciding in advance what the solutions should be. Control all conversations. Ask only closed *yes* or *no* questions. Make working together feel like drudgery.

In the process of developing Liberating Structures and exposing students and employees in many different countries and environments, we have come to the conclusion that: You can't get to a "good life" if you don't know *how to* do it. So here is our proposition: Use routinely a collection of simple methods called Liberating Structures and your chances for a "good life" for you and those around you, at school and at work, will be dramatically increased.

REFERENCES

Bruner, J. (1960). *The process of education.* Cambridge, MA: Harvard University Press.
Bruner, J. (1973). *Beyond the information given.* New York: Norton.
Bruner, J. (1996). *The culture of education.* Cambridge, MA: Harvard University Press.
Darling-Hammond, L. (2013). *Powerful teacher education: Lessons from exemplary programs.* San Francisco, CA: Jossey-Bass.
Davis, J. (2013, October 15). How a radical new teaching method could unleash a generation of geniuses. *Wired.* Retrieved from http://www.wired.com/business/2013/10/free-thinkers/
Dewey, J. (1938). *Experience and education.* New York: Macmillan.
Freire, P. (1971). *Pedagogy of the oppressed.* (M. B. Ramos, Trans.). New York: Continuum. (Original work published 1968)
Kolb, D. A. (1984). *Experiential learning: Experience as the source of learning and development.* Upper Saddle River, NJ: Prentice-Hall.

Lipmanowicz, H., & McCandless, K. (2014). *The surprising power of liberating structures: Simple rules to unleash a culture of innovation.* Scotts Valley, CA: CreateSpace. Available from https://www.createspace.com/4419178

Montessori, M. (1986). *Discovery of the child.* New York: Ballantine.

Papa, M. J., Singhal, A., & Papa, W. H. (2006). *Organizing for social change: A dialectic journey of theory and praxis.* New Delhi: Sage.

Piaget, J. (2001). *The psychology of intelligence.* (M. Piercy, Trans.). New York: Routledge. (Original work published 1947)

Singhal, A. (2014). Creating more substance, connections, and ideas in the classroom. In H. Lipmanowicz & K. McCandless (authors), *The surprising power of liberating structures: Simple rules to unleash a culture of innovation.* Scotts Valley, CA: CreateSpace. Available from https://www.createspace.com/4419178

Singhal, A., Buscell, P., & Lindberg, C. (2014). *Inspiring change and saving lives: The positive deviance way.* Bordentown, NJ: Plexus.

Singhal, A., McCandless, K., Buscell, P., & Lindberg, C. (2009). Spanning silos and spurring conversations: Positive deviance for reducing infection levels in hospitals. *Performance, 2*(3): 78–83.

ADDITIONAL RESOURCES FOR LIBERATING STRUCTURES

Liberating classrooms: Henri Lipmanowicz in conversation with Dr. Arvind Singhal. [Video file]. Retrieved from https://vimeo.com/50347352 (8' 20")

Liberating structures: Including and unleashing ALL: Henri Lipmanowicz in conversation with Dr. Arvind Singhal. [Video file]. Retrieved from https://vimeo.com/50352840 (8'30")

Liberating structures: Simple, subtle, powerful. [Video file]. Retrieved from http://clintonschool-speakers.com/content/liberating-structures-simple-subtle-powerful

Nine liberating structures in 42 minutes with Henri Lipmanowicz and Dr. Arvind Singhal. [Video file]. Retrieved from http://vimeo.com/60843778

Unscripted: Liberating structures by Dr. Arvind Singhal. [Video file]. Retrieved from https://vimeo.com/51546509 (10'20")

Contributors

Saleem Alhabash is an Assistant Professor of Public Relations and Social Media, jointly appointed by the Department of Advertising + Public Relations and the Department of Media and Information at Michigan State University. He received his PhD from the University of Missouri's School of Journalism. His research focuses on the processes and effects of using new and social media. More specifically, his research untangles the ways in which computer-mediated communication can facilitate persuasion.

Meryl Alper is a PhD candidate in Communication at the Annenberg School for Communication and Journalism at the University of Southern California. She studies the social and cultural implications of networked communication technologies, with a particular focus on disability and digital media, children and families' technology use, and mobile communication. Prior to USC, she worked in the children's media industry as a researcher and strategist with Sesame Workshop, Nickelodeon, and Disney. Her research has been published in a number of journals, including *New Media & Society*, *International Journal of Communication*, and *Journal of Early Childhood Literacy*. Her book, *Digital Youth with Disabilities* (MIT, 2014), examines the out-of-school media and technology experiences of children, adolescents, and teenagers with disabilities in the digital age. Alper graduated magna cum laude from Northwestern

University, double majoring in Communication Studies and History. She also holds a certificate in Early Childhood Education from UCLA.

Peng Hwa Ang is a Professor at the Wee Kim Wee School of Communication and Information, Nanyang Technological University, Singapore, and President-Elect-Select 2014 of the ICA. A lawyer by training, he worked as a journalist before going on to pursue an MA in communication management at the University of Southern California and a PhD in mass media at Michigan State University. His teaching and research interests combine law and communication, touching on Internet governance, censorship, and the social impact of media. He was a member of the UN Working Group on Internet Governance, later going on to co-found the Global Internet Governance Academic Network, a community of academics researching Internet governance, and served as its inaugural chair. He also co-founded the Asia Pacific Regional Internet Governance Forum, and served as the inaugural chair.

David Atkin is a Professor at the University of Connecticut. He studies the uses and effects of new media—with a particular focus on the diffusion and uses of emerging media—including uses of the social media for health information. He received his PhD from Michigan State University. In 1999, Dr. Atkin received the Kreighbaum Under 40 Award, recognizing a top junior journal scholar in the field, and in 2000, he received a University Distinguished Scholar Award. He has taught courses on communication technology, campaigns, policy, and criticism. Dr. Atkin has published four books and 130 articles, ranking one of the 100 most prolific scholars (top 1%) in the history of the discipline. He has worked on research grants and contracts supported by such agencies as Ameritech Foundation, National Association of Broadcasters, SHB law firm and various university sources. Dr. Atkin is Associate Editor at *JQMC Quarterly* and also serves on several journal editorial boards.

Heli Tuomi Carlile is a Creative Producer and Senior Project Manager at Parallel World Labs. She has over 20 years' experience in a wide range of cross-platform media, including interactive, online, film, print, and broadcast. Prior to joining Parallel World Labs in 2003, she directed the production division at a creative agency in New York where her Fortune 500 clients included FedEx, Sony, and Fidelity. In her role at Parallel World Labs, she has actively contributed to the realization of various projects in experience centers, museums and cultural institutions, aquariums, world expositions, and the Olympic Games. She holds a degree in Film from Queen's University, an MA in Journalism from University of Western Ontario, and has also studied at the Cyber-Architecture Unit / McLuhan Program at University of Toronto, and the Graduate School of Design at Harvard. Her projects with Parallel World Labs have received numerous awards including the Game Pioneer Award and a Producer of the Year nomination from the Canadian New Media Awards, and an Applied Arts award for an interactive restaurant in Vancouver.

Shelia R. Cotten, a sociologist, is a Professor in the Department of Media and Information at Michigan State University. She studies technology use across the life course, and the social, educational, and health outcomes of using various technologies. Her research has been funded by the National Institutes of Health and the National Science Foundation. Her work has been recently published in *Computers & Education, Social Science & Medicine, Computers in Human Behavior, Journal of Family Issues, Journal of Applied Gerontology, Journal of Gerontology: Social Sciences,* and *Information, Communication, and Society.* She and Laura Robinson are the co-editors of the Emerald Series in Media and Communication. In 2013, she won the award for Public Sociology from the Communication and Information Technologies section of the American Sociological Association (CITASA).

Dimitrina Dimitrova is a Lecturer at York University, Toronto, and the founding Project Director of the NAVEL/ENOW study at the University of Toronto's NetLab. She has extensive experience in studying telework and networked work. She is the lead author of "NAVEL: Studying a Networked Organization" (2013), "Managing Collaborative Research Communities" (2010), and "Research Communities in Context" (2009).

Allison L. Eden is an Assistant Professor in the Department of Communication Science at the VU University Amsterdam. She holds a PhD in Communication from Michigan State University. Her research focuses on understanding media entertainment and enjoyment processes and effects from a media psychological perspective. Her current work is split between two main areas. The first connects morality to media enjoyment and selection, using morality as both a predictor and an outcome variable in examining entertainment. The second area is the motivational role of enjoyment in narrative and video games, focusing on the potential of entertainment's ability to promote personal growth and pro-social behavior.

Elizabeth Ellcessor is an Assistant Professor of Cinema and Media Studies at Indiana University, Bloomington, where she is also affiliated faculty in Gender Studies and Cultural Studies. Her research on disability and digital media access has been published in *Information, Communication & Society, Television & New Media,* and *Critical Studies in Media Communication.* She also serves on the editorial board of *Disability Studies Quarterly.* Additional research interests include media industry studies and digital media celebrity; this work has been published in *Cinema Journal* and appears in an anthology, *Making Media Work* (NYU Press, 2014).

Katie Ellis is a Senior Research Fellow in the Internet Studies Department and member of the Centre of Culture and Technology at Curtin University. She holds an Australian Research Council (ARC) Discovery Early Career Researcher Award studying disability and digital television. She has worked with people with disability in the community, government, and in academia and has published widely in the area of disability and digital and networked media. Her books include *Disabling Diversity* (VDM Verlag, 2008), *Disability and New Media* (Routledge, 2011; with Mike Kent), *Disability, Obesity and Ageing: Popular Media Identifications* (Ashgate, 2014; with Debbie Rodan and Pia Lebeck), *Disability and the Media* (Palgrave, 2015, with Gerard Goggin) and *Disability and Popular Culture: Focusing Passion, Creating Community, Expressing Defiance* (Ashgate, 2015).

Charles M. Ess is a Professor in Media Studies, Department of Media and Communication, University of Oslo. Ess has received awards for excellence in teaching and scholarship; he has also held several guest professorships in Europe and Scandinavia, most recently as Guest Professor, Philosophy Department, University of Vienna (2013–2014). Ess has published extensively in Information and Computing Ethics (e.g., *Digital Media Ethics*, 2nd ed., Polity, 2013) and in Internet Studies (e.g., with William Dutton, "The Rise of Internet Studies," *New Media & Society*, 15(5), 2013). Ess emphasizes cross-cultural approaches to media, communication, and ethics, focusing especially on virtue ethics and its illuminations of being human in an (analogue-)digital age.

Gerard Goggin is an Australian Research Council Future Fellow and a Professor of Media and Communications at the University of Sydney. He has published widely on digital technology, his books including *Routledge Companion to Global Internet Histories* (2016; with Mark McLelland), *Routledge Companion to Mobile Media* (2014; with Larissa Hjorth), *Technologies and the Media* (2013), *Global Mobile Media* (2011), *Cell Phone Culture* (2006), and the co-edited collections *Locative Media* (2014), *Mobile Technology and Place* (2012), *Internationalizing Internet Studies* (2009), and *Mobile Technology*. In addition, he is well-known for his work on disability and media, including, with Katie Ellis, the books *Routledge Companion to Disability and Media* (2017; also with Beth Haller) and *Disability and the Media* (2015), and, with the late Christopher Newell, *Disability in Australia* (2005), and *Digital Disability* (2003).

Maren Hartmann is an Assistant Professor at the University of the Arts Berlin (UdK). She received her PhD from the University of Westminster and has worked at several universities in the UK, Belgium, and Germany in both research and teaching positions. Until recently, she was a member of the Executive Board of ECREA as well as (Vice-) Chair of the media sociology section of the DGPuK. She also founded (and led for some time) the digital culture and communication section of ECREA. Her research focuses mostly on media appropriation processes and concepts (esp. domestication), on digital media cultures, on the relationship between media and space(s), on the question of the materiality of media, and on mobility. She has published widely in these fields. A recent article appeared in the first issue of *Mobile Media and Communication*.

Tilo Hartmann is an Associate Professor at the Department of Communication Science at the VU University Amsterdam. He holds a PhD in Communication Science from the University of Music, Drama and Media Hannover. In his research he applies media psychological approaches and methodology to study people's experience of mediated illusions (e.g., parasocial interaction, presence, virtual violence), media choice, and health communication behavior. Tilo Hartmann is editor of *Media Choice: A Theoretical and Empirical Overview*, and editorial board member of *Journal of Communication*, *Human Communication Research*, and *Media Psychology*.

Andreas Hepp is a Professor of Media and Communication Studies at the ZeM-KI, Centre for Communications, Media and Information Research, University of Bremen, Germany. He is co-initiator and principal investigator of the DFG priority program "Mediatized Worlds" and the Creative Unit "Communicative Figurations" (a research network with the University of Hamburg). Hepp's main research areas are media and communication theory, media sociology, mediatization research, transnational and transcultural communication, media change, and methods of media culture research. His publications include *Media Events in a Global Age* (Routledge, 2010; co-editors N. Couldry and F. Krotz), *Cultures of Mediatization* (Polity, 2013), and *Mediatized Worlds* (Palgrave, 2014; co-editor F. Krotz).

Leo W. Jeffres is a Professor Emeritus of Communication at Cleveland State University. He received his PhD from University of Minnesota in 1976. A contributor to most of the major streams of research in mass communication over the past 35 years, he is the author of more than a hundred journal articles and four books, including *Mass Media Processes* (1994), *Mass Media Effects* (1997), and *Urban Communication Systems: Neighborhoods and the Search for Community* (2002). His interests include neighborhoods and urban communication systems, communication technologies, media effects, and ethnic communication. Recently, he advanced "communication capital" as a concept distinct from "social capital," and he has developed the "community/urban communication audit" as a diagnostic tool for assessing the wide range of communication assets available in a geographic community. He is a former Fulbright scholar, a MAPOR Fellow, and a member of the Urban Communication Foundation's board of directors.

Mengtian Jiang is a PhD student in the Media and Information Studies program with a concentration in consumer psychology at Michigan State University. Prior to joining the PhD program, she earned her MA in Advertising and her BA in Journalism. She is interested in understanding how people are influenced and persuaded by online information and online strangers, with a particular focus on consumer trust.

Susan B. Kretchmer is Co-Founder and President of Partnership for Progress on the Digital Divide, a not-for-profit organization engaging a broad diversity of individuals and organizations to spearhead a multi-associational, multidisciplinary partnership between scholars, practitioners, and policymakers to make significant contributions in closing the digital divide and addressing other challenges and opportunities of the digital age. Since 1986, Kretchmer has been active in service and professional organizations in Communication and authored over 100 papers, journal articles, and book chapters, numerous "Top Five" papers and convention keynotes, and is editor of the forthcoming book, *Navigating the Network Society: The Challenges and Opportunities of the Digital Age*. Her research explores the historical, social, and cultural relationship between communication and technology in popular media, law and public policy, and social change, with special expertise in issues of marginalized populations and ICTs. In addition, Kretchmer has advised the U.S. Federal Communications Commission, the U.S. Supreme Court, and the government of Canada.

Robert LaRose is a Professor in the Department of Media and Information at Michigan State University where he teaches graduate courses in research methods and theory and serves as Director of the Media and Information Studies PhD program. His research interests are the uses and effects of new media. His current foci are the role of habits in media behavior and the adoption of broadband Internet among vulnerable populations. He is the co-author of a popular introductory textbook, *Media Now*. He was presented with the Outstanding Article Award for 2011 by the International Communication Association. He holds a PhD in Communication Theory and Research from the Annenberg School at the University of Southern California.

Rich Ling is the Shaw Foundation Professor of Media Technology at Nanyang Technological University, Singapore. He also has connections with Telenor Research and has an adjunct position at the University of Michigan. Ling holds a PhD in Sociology from University of Colorado and has studied the social consequences of mobile communication for the past two decades. He has written *The Mobile Connection* (Morgan Kaufmann, 2004), *New Tech, New Ties* (MIT, 2008) and most recently *Taken for Grantedness* (MIT, 2012). He is a founding co-editor of *Mobile Media and Communication* (Sage) and the Oxford University Press series Studies in Mobile Communication.

Henri Lipmanowicz is co-developer of Liberating Structures Press and co-author of *The Surprising Power of Liberating Structures: Simple Rules to Unleash a Culture of Innovation*. He was Chairman and co-founder of Plexus Institute from 2000 to 2011. Prior to that, he had 30 years at Merck & Co. progressing from Managing Director in Finland to President of the Intercontinental Division and Japan from 1967 to 1997. He holds an MS in Industrial Engineering from Columbia University, New York and an MS in Chemical Engineering from Université de Toulouse, France.

Peter Lunt is a Professor and the Head of the Department of Media and Communication at the University of Leicester, UK. Trained as a social psychologist, his main areas of research have been in the psychology of consumption, media audiences (particularly public participation in and through popular culture) and, more recently, media regulation. His recent books are *Stanley Milgram* (Palgrave, 2011) and *Media Regulation* (Sage, 2012; with Sonia Livingstone). He is currently writing a book *Goffman and the Media* for Polity with Espen Ytreberg and conducting a project on media portrayals of the relation between moral and political discourse in discussions of social justice with David Scott.

Keith McCandless is a founding partner of the Social Invention Group in Seattle, a consultant with expertise in innovating by including and unleashing everyone, co-developer of Liberating Structures Press, and co-author of *The Surprising Power of Liberating Structures: Simple Rules to Unleash a Culture of Innovation*. He served on the Board of Directors and Science Advisory Board of the Plexus Institute and has a master's degree from the Heller School at Brandeis University in Boston.

Diana Mok is an Associate Professor at the University of Western Ontario with a joint appointment in Management and Organizational Studies and Geography. A member of the NetLab Network, her primary research interests include urban real estate economics and applications of geographic information sciences, with a focus on housing. Her interests have recently broadened to look at the spatiality of social networks.

Kimberly A. Neuendorf is a Professor in the School of Communication at Cleveland State University. She received her PhD from Michigan State University in 1982. A prolific researcher, her interests include media and race/ethnicity, new media technologies, and affective correlates of media use. Her recent work has pioneered investigations on the "senses of humor" and media use, and she currently is studying diverse aspects of film, including how audiences process films as messages. Her book, *The Content Analysis Guidebook* (2002) is used widely by scholars across disciplines.

Mary Beth Oliver is a Distinguished Professor at Pennsylvania State University in the Department of Film/Video & Media Studies and co-director of the Media Effects Research Lab. Her research focuses on entertainment psychology and on social cognition and the media. Her recent publications have appeared in such journals as the *Journal of Communication, Human Communication Research*, and *Communication Research*, among others. She is currently an associate editor of the *Journal of Media Psychology* and former editor of *Media Psychology* and associate editor of the *Journal of Communication* and *Communication Theory*. She is co-editor on several books, including *Media and Social Life, Media Effects: Advances in Theory and Research*, and *The Sage Handbook of Media Processes and Effects*. She was elected in 2014 as a fellow of the International Communication Association, and is currently working on a co-edited handbook on media and well-being and positive media psychology.

Joy Pierce is an Assistant Professor in the Department of Communication at the University of Utah. She holds a PhD from University of Illinois at Urbana-Champaign. Her work focuses on new media technologies and underserved populations, particularly Black and Mexican immigrant populations. Pierce employs critical cultural and contemporary social theories as well as qualitative research methods to interrogate inequities in digital media use among historically marginalized populations. Her works in new and emerging technologies, information communication policy, and social problems have been presented at national and international conferences. She has published in communication, sociology, and qualitative methods journals and served as guest editor for a special issue of *Social Identities*. Pierce is completing a book manuscript that articulates how digital literacy programs may compete, conflict, and empower the everyday lives of two communities as they attempt to become part of the mainstream network society.

Leonard Reinecke is an Assistant Professor at the Department of Communication at Johannes Gutenberg University Mainz, Germany. He holds an MA and a PhD in Psychology from the University of Hamburg. His work has addressed the uses and effects of interactive and non-interactive media, computer-mediated communication, and entertainment media. His current research is focused on the interactions of media use and well-being. His work has investigated the potential of entertainment media to facilitate recovery from stress and strain, the satisfaction of intrinsic psychological needs through the use of video games, and the implications of online self-disclosure and online social support for psychological well-being. Most recently, his research has addressed the moderating role of appraisal processes and self-control for the effects of media use on well-being.

Nora J. Rifon is a Professor in the Department of Advertising + Public Relations at Michigan State University. She earned her PhD in Business, and her MA and BA in Psychology. Her research interests include consumer privacy and online safety, marketing communications strategies, corporate reputation, sponsorship, and children and media. Her work has been published in *Communications of the ACM*, *New Media & Society*, *The Journal of Consumer Affairs*, *The Journal of Advertising*, *Advances in Consumer Research*, *Government Information Quarterly*, *The Journal of Interactive Advertising*, and *The International Journal of Advertising*, and in the proceedings of a variety of international conferences. She has served on the Executive Committee and the Publications Committee of the American Academy of Advertising, and on the editorial review boards of *The Journal of Advertising*, *The Journal of Consumer Affairs*, and *The Journal of Interactive Advertising*, and served as consultant to the State of Michigan Office of the Attorney General, private law firms, and the commercial sector.

Laura Robinson is an Assistant Professor in the Department of Sociology at Santa Clara University. She earned her PhD from UCLA, where she held a Mellon Fellowship in Latin American Studies and received a Bourse d'Accueil at the École Normale Supérieure. In addition to holding a postdoctoral fellowship on a John D. and Catherine T. MacArthur Foundation funded project at the USC Annenberg Center, Robinson has served as Visiting Assistant Professor at Cornell University and Visiting Scholar at Trinity College, Dublin. Her research has earned awards from CITASA, AoIR, and NCA IICD. Robinson's current multi-year study examines digital and informational inequalities. Her other publications explore interaction and identity work, as well as new media in Brazil, France, and the United States. Her website is www.laurarobinson.org. Email: laura@laurarobinson.org.

Jian Raymond Rui is a Visiting Assistant Professor in the Department of Communication Studies, Texas Tech University. He received his PhD from University at Buffalo, The State University of New York. His research interest is how people use new communication technology and how new technology interacts with social, cultural, relational, and psychological factors to affect the outcome of interpersonal communication. His work has been published in *Information, Communication & Society, CyberPsychology, Behavior, and Social Networking*, and *Computers in Human Behavior*. His "The Relationship Between on- and Offline Social Behavior, and Social Capital" received the top student paper award in the Human Communication & Technology Division at the 99th Annual Conference of National Communication Association (NCA-99, 2013).

Ruth Shillair is a PhD student in the Media and Information Studies program at Michigan State University. Her background includes management, computer information management, teaching, and working with student populations from a wide range of backgrounds. Her research interests include understanding the characteristics of generative learning communities and finding ways to equip these communities to improve diffusion of information to diverse populations. An important part of this process is to understand how different age cohorts interact online, protect themselves, and build trust.

Arvind Singhal is the Samuel Shirley and Edna Holt Marston Endowed Professor of Communication and Director of the Social Justice Initiative in the University of Texas at El Paso's Department of Communication. He also serves (since 2009–2010) as the William J. Clinton Distinguished Fellow at the Clinton School of Public Service, University of Arkansas at Little Rock, Arkansas. Singhal teaches and conducts research in the diffusion of innovations, the positive deviance approach, organizing for social change, the entertainment-education strategy, and liberating structures. His research and outreach span sectors such as health, education, sustainable development, civic participation, and corporate citizenship. Singhal is co-author or editor of 13 books, including *Inspiring Change and Saving Lives: The Positive Deviance Way* (2014); *Health Communication in the 21st Century* (2014); *Inviting Everyone: Healing Healthcare through Positive Deviance* (2010); *Protecting Children from Exploitation and Trafficking: Using the Positive Deviance Approach* (2009); *Communication of Innovations* (2006); *Organizing for Social Change* (2006); and *Entertainment-Education: A Communication Strategy for Social Change*

(1999). In addition, Singhal has authored some 175 peer-reviewed essays in journals of communication, public health, and social change and won over two dozen international and national awards. He has visited and lectured in over 70 countries of Asia, Africa, Latin America, Australia, Europe, and North America.

Stacey Spiegel is the CEO, Creative Director, and Co-founder of Parallel World Labs and a new media artist working internationally in the design and development of advanced media experiences. Under Spiegel's leadership, Parallel World Labs has worked with prestigious clients such as The Smithsonian Institution, The Science Museum in London, Cité des Sciences de Paris, the New England Aquarium, Bob Ballard Institute for Exploration, NOAA National Oceanic & Atmospheric Administration, the government of Canada, the government of Norway, and the International Olympic Committee (IOC). Rockheim, Norway's National Museum of Rock and Pop, designed by Spiegel, has been widely praised by media and was nominated for the 2012 European Museum of the Year. A former fellow of MIT Center for Advanced Visual Study in Boston, Spiegel has served as Adjunct Professor of Architecture and Landscape Architecture at the University of Toronto and Adjunct Professor in the Department of Software Engineering and Game Design at McMaster University.

Michael A. Stefanone is an Associate Professor in the Department of Communication, University at Buffalo, The State University of New York. He holds a PhD from Cornell University. His research interest focuses on the social psychology of technology use, specifically, the effects of technology on human relationships and access to resources like social capital. He has published in *Journal of Computer-Mediated Communication, CyberPsychology, Behavior, and Social Networking*, and *Social Science Computer Review*. His "Do Me a Solid? Information Asymmetry, Liking, and Compliance Gaining Online" received the best paper award in the Social Networks minitrack, Organizational Systems and Technology track, at the 45th Annual Hawaii International Conference on Systems Science (HICSS-45, 2012).

Hsin-yi Sandy Tsai is a PhD candidate in the Department of Media and Information at Michigan State University. Her research interests are communications and technologies, including telecommunication policies, technology adoption and use, public media, civic engagement, and digital inclusion. She is especially interested in how to make better policies to help people improve their quality of life and take full advantage of new technologies.

Peter Vorderer is a Professor of Media and Communication Studies at the University of Mannheim in Germany. Previous affiliations include the University of Music, Theater and the Media in Hannover, the Annenberg School for Communication at the University of Southern California and the Free University of Amsterdam. His research focuses on users' interest in entertainment and in new media, as well as on consequences of using them. Currently, he is particularly interested in people's habit to be (almost) permanently online and connected with others. His research has been published in major communication journals and in 11 authored and edited books. He currently serves as President of the International Communication Association.

Hua Wang is an Assistant Professor in the Department of Communication and Research Assistant Professor in the Department of Community Health and Health Behavior at the University at Buffalo, The State University of New York. She holds a PhD from the Annenberg School for Communication & Journalism at the University of Southern California. Her research includes communication technologies, social networks, health promotion, and social change. Her work has been published in *Communication Research, American Behavioral Scientist, Computers in Human Behavior, CyberPsychology, Behavior, and Social Networking,* among others. Her most recent projects focus on using innovative approaches such as transmedia storytelling, social impact games, positive deviance, and liberating structures to address complex public health and social issues.

Barry Wellman is a Professor Emeritus of Sociology and a member of the NetLab Network at the iSchool, University of Toronto. He is the Principal Investigator of the NAVEL/ENOW project. Wellman co-authored *Networked: The New Social Operating System.* He founded the International Network for Social Network Analysis in 1976, is a Fellow of the Royal Society of Canada, and has won multiple career achievement awards.

Julia K. Woolley is an Assistant Professor of Communication Studies at California Polytechnic State University, San Luis Obispo. She received her PhD from Pennsylvania State University. Her research examines media processes and effects in the areas of entertainment psychology and new communication technologies. Recent publications include co-authored articles appearing in *Mass Communication and Society* and *Computers in Human Behavior.*

Author Index

Subject Index

international
communication
association

Annual Conference Theme Book Series

As of 2013, ICA and Peter Lang Publishing started co-publishing papers from ICA's annual conference theme sessions in the form of edited collections. Written in an engaging style, these volumes are meant to appeal to a wider audience and to reach scholars in other disciplines outside of Communication Studies. As such, the collections are not conference proceedings per se but a unique set of selected essays that capture the insights and agendas of the discipline's top scholars.

For additional information about this series, please contact Mary Savigar at Mary.Savigar@plang.com.

To order other books in this series, please contact our Customer Service Department:

(800) 770-LANG (within the U.S.)
(212) 647-7706 (outside the U.S.)
(212) 647-7707 FAX

Or browse online by series at www.peterlang.com

www.ingramcontent.com/pod-product-compliance
Lightning Source LLC
Chambersburg PA
CBHW070938050326
40689CB00014B/3252